An Introduction to
TEXTILE MECHANISMS

AN INTRODUCTION TO TEXTILE MECHANISMS

P. GROSBERG

Research Professor of Textile Engineering
The University at Leeds

ERNEST BENN LIMITED

LONDON · 1968

First published 1968 by Ernest Benn Limited
Bouverie House · Fleet Street · London · EC4
© P. Grosberg 1968
Distributed in Canada by
The General Publishing Company Limited · Toronto
Printed in Great Britain

510–46101–8

PREFACE

A CONSIDERABLE part of the training of all textile technologists is concerned with the detailed study of the numerous mechanisms used in the processing of textile materials. The great majority of textile students, however, have no background knowledge of engineering principles and, as a result, are unable to appreciate either the finer points of the mechanisms studied or the connection between these mechanisms and the totality of available mechanisms. For several years now the author has been presenting to textile students a special course in textile mechanisms with the aim of giving the student a basis for making a comparative study of textile mechanisms. This book is intended to make this approach available to a wider audience.

While, for obvious reasons, the book follows in broad outlines the many available texts in theory of machines, it differs from these in two important respects. Firstly, the balance of the topics discussed is biased by the particular types of mechanisms used in the textile industry. Thus, for example, cams and selection mechanisms are considered in much greater detail than is usual in elementary texts. In addition, the examples used to illustrate the principles discussed are taken exclusively from the textile machines with which the student is familiar. Secondly, the approach used is geared to the basic mathematical engineering knowledge that textile students usually possess. It is found that, in general, textile students are unable to appreciate the generalisations of kinematics without a careful discussion of some of the particular mechanisms which have led to these generalisations. As a result, the analytical methods used are not in general rigorous, but are aimed mainly at helping the student to appreciate that the mechanisms he is studying are only part of the general study of machinery. It is hoped that this approach will encourage him to read the more specialised books which are available on the subject, and references to such works are included after each chapter.

The book has been aimed in general at the ordinary and honours degree student in textile technology, as well as the student sitting for his higher national certificate or diploma in textile technology. Such a student should already possess some knowledge of textile processing and, in addition, be acquainted with the basic principles of differential and integral calculus.

I am deeply indebted to three of my colleagues for their help; to Mr G. A. V. Leaf for reading the original draft and suggesting many simpler and neater methods of presentation of the material; to Dr K. Hepworth for suggesting the novel method for describing the behaviour of selection mechanisms used in Chapter 6, and to Mrs B. McClelland both for preparing the final text for publication and for her gentle chiding which ensured that the book was finally completed. I am also indebted to the many firms and publishers who have allowed me to reproduce diagrams and photographs.

P. G.

The University at Leeds
November, 1967

CONTENTS

PREFACE · · · *page* v

1 RESTRAINTS, FRICTION AND LUBRICATION · · · 1

    1.0 Introduction · 1.1 The nature of friction · 1.2 Lubrication by means of chemical films · 1.3 Lubrication by thick films of fluid · 1.4 The use of rolling instead of sliding contact: 1.5 Journal and roller bearings · 1.6 The design of sliding pairs

2 THE TRANSMISSION AND MODIFICATION OF ROTATIONAL MOTION —FRICTIONAL DRIVES · · · 19

    2.0 Introduction · 2.1 Definition of rotational speed · 2.2 Frictional drives between two parallel shafts · 2.3 Belt drives · 2.4 Conditions of critical slipping of belts · 2.5 Variable speed friction drives

3 THE TRANSMISSION AND MODIFICATION OF ROTATIONAL MOTION —POSITIVE DRIVES · · · 38

    3.0 Introduction · 3.1 Roller chain drives · 3.2 Timing belts · 3.3 Gear drives · 3.4 Spur gearing · 3.5 The shape of gear teeth · 3.6 The involute gear · 3.7 The gear tooth height · 3.8 The efficiency of spur gears · 3.9 Worm gears · 3.10 The efficiency of worm gears · 3.11 Simple gear trains 3.12 Epicyclic or Planetary gear trains · 3.13 The differential gear · 3.14 Further uses of epicyclic gear trains · 3.15 The efficiency and choice of gear trains

4 CAMS · · · 75

    4.0 Introduction · 4.1 Cams · 4.2 Motion of the cam follower · 4.3 The linear cam · 4.4 The parabolic cam · 4.5 The forces on the cam disc · 4.6 The effect of friction on the cam load · 4.7 The pressure angle on a roller follower · 4.8 The vibration of the cam follower · 4.9 The effect of linkage elasticity on the follower movement · 4.10 The use of positive cams

5 LINKAGE MECHANISMS · · · 107

    5.0 Introduction · 5.1 The four-bar linkage · 5.2 Simple harmonic motion · 5.3 Modified simple harmonic motion · 5.4 Multiple-bar linkages · 5.5 Combined ratchet and linkage mechanisms · 5.6 Examples of complex combined mechanisms

6 SELECTION MECHANISMS · · · 131

    6.0 Introduction · 6.1 Methods of storing information · 6.2 The grouping of machine parts for selection · 6.3 Converting information into movement · 6.4 Some mechanical switching mechanisms · 6.5 Dobby selection mechanisms · 6.6 High speed mechanical switching mechanisms · 6.7 Mixed mechano-electrical switching mechanisms · 6.8 Some more complex mechanical switches · 6.9 The movement of the information store

7 CONTROL MECHANISMS 172

7.0 Introduction · 7.1 The elements of a control mechanism · 7.2 A comparison of open- and closed-loop systems · 7.3 The detecting element · 7.4 The detection of broken ends of yarns and slivers · 7.5 The control of yarn and cloth tension · 7.6 Let-off and take-up mechanisms · 7.7 The control of sliver thickness in drafting and carding 7.8 The detection of full or empty packages and their replacement 7.9 The control of temperature

INDEX 197

LIST OF FIGURES

CHAPTER 1

1.1 Picking stick in loom, showing various parts *page* 2
1.2 Flow of lubricant between two inclined plane surfaces 5
1.3 Contact between rolling body and a plane 6
1.4 Restraints on a shaft 8
1.5 Position of shaft in journal bearing under differing loads and at differing speeds 8
1.6 Experimental results for friction in a journal bearing, obtained by Prof. A. Tenot [*Courtesy of* Inst. of Mechl. Eng.] [*Reproduced from* 'Proceedings of Inst. of Mechl. Engineers: General Discussion on Lubrication', vol. 1, p. 321, 1936] 9
1.7 f versus NZ/W. 10
1.8 Hydrostatic bearing [*Reproduced from* Morrison and Crossland 'An Introduction to the Mechanics of Machines', p. 445, Longmans, 1964] 11
1.9 Magnetic bearing 12
1.10 a. Deep groove ball; b. roller; c. thrust; d. cylinder, and e. self-aligning ball bearing [*Courtesy of* S.K.F. Ltd] 13
1.11 Effect of wear on: a. a short bearing, b. a long bearing, and c. a two-bearing support 14
1.12 Position of shaft under load with fixed and self-aligning bearings 15
1.13 Ring spinning bearing unit [*Courtesy of* S.K.F. Ltd] 16
1.14 Sliding pair using journal bearings 17
1.15 Linear ball bearing [*Courtesy of* Rotalin Bearings Ltd] 17

CHAPTER 2

2.1 The transmission of rotational motion between two discs 20
2.2 Cone winder [*Courtesy of* Schweiter Engineering Works Ltd] 21
2.3 Loading system for drafting rollers [*Courtesy of* Prince-Smith and Stells Ltd] 22
2.4 Loading system on pressure mangle [*Courtesy of* Sir James Farmer Norton & Co. Ltd] 23
2.5 Relative motion of pulleys in a belt drive 24
2.6 Tape drive to spindles 24
2.7 Forces on a small segment of a belt 26
2.8 Forces on the cross-section of a V-belt 27
2.9 Dimensions of a variable-speed belt drive 30
2.10 Cone pulleys on flyer drawing frame 32

2.11 Cone pulleys on Raper autoleveller
  [Courtesy of Prince-Smith and Stells Ltd]   33
2.12 Positive feed cone for circular weft knitting machine
  [Courtesy of Hosiery and Allied Trades Research Association]   34
2.13 Cone pulleys on warp beam speed control of FNF warp knitting machine   35
2.14 Plate and wheel variable-speed drive   35
2.15 Variable-speed drive of Kopp Variator   *[Courtesy of* Allspeeds Ltd]   36
2.16 Variable-speed V-belt drive
  [Courtesy of U.S. Electrical Motors Inc.]   36
2.17 The PIV variable-speed drive
  [Courtesy of the Link Belt Company]   37

CHAPTER 3

3.1 Chain drive   38
3.2 Chain drive to intermediate rollers on drafting frame
  [Courtesy of Prince-Smith and Stells Ltd]   39
3.3 Chain drive to carding workers and strippers   39
3.4 Chain drive in two positions of engagement   40
3.5 Timing belt   *[Courtesy of* Crofts (Engineers) Ltd]   41
3.6 Various forms of gear drives
  [Courtesy of David Brown Corporation (Sales) Ltd]   42
3.7 Principal dimensions of two meshing gears   44
3.8 Velocities of contact point of two gears   47
3.9 Properties of involute   49
3.10 Production of involute teeth—general view   50
3.11 Production of first involute tooth   50
3.12 Production of mating involute tooth   50
3.13 Line of pressure between two involute teeth   51
3.14 The effect of tooth size on meshing length   54
3.15 Cycloidal gear—Roots blower   55
3.16 Contact velocities on involute gear   56
3.17 Part-sectioned view of worm gear
  [Courtesy of David Brown Corporation (Sales) Ltd]   57
3.18 The helix and its properties   57
3.19 Forces acting on worm and wheel   59
3.20 Efficiency of worm gears for varying coefficients of friction and helix angle   60
3.21 Simple gear trains   61
3.22 Headstock gear trains   63
3.23 Planetary gear train   64
3.24 Simple analogous gear train   65
3.25 Planetary gear train   67
3.26 Differential gear   69
3.27 Gear train for cone-flyer variable-speed drive   70

3.28 Epicyclic gear for sliver feed
                            [*Courtesy of* Carding Specialists Ltd]   71
3.29 Required feeding motion   72
3.30 Front draft-twist rollers from woollen drafting frame   72
3.31 Variation in speed of take-up roller
                    [*Courtesy of* Wool Industries Research Association] 73
3.32 Preferred gear design for the take-up roller   74

CHAPTER 4

4.1   Disc cam with knife-edged follower   76
4.2   End cam   76
4.3   a. Knitting cam; b. mushroom follower and disc cam; c. roller follower 77
4.4   Cams used to control movements of the parts of a flat-bed knitting
      machine   77
4.5   Cylindrical cam   78
4.6   Face groove cam   78
4.7   Conjugate cams [*Reproduced from* H. A. Rothbart *Cams—Design,
      Dynamics and Accuracy*, Wiley, 1956]   79
4.8   Cam which will produce a simple harmonic motion   80
4.9   Linear or heart-shaped cam   81
4.10 Displacement, velocity and acceleration of the follower with a linear
      cam   82
4.11 Displacement, velocity and acceleration of the follower with a para-
      bolic cam   85
4.12 Forces acting on a knitting needle   91
4.13 Path of yarn in a knitting machine   93
4.14 Pressure angle of cam disc and roller   94
4.15 Vibration caused by high-speed running with a. parabolic; b. har-
      monic, and c. cycloidal cams
      [*Reproduced from* D. B. Mitchell, *Mech. Eng.* **72**, p. 467, June 1950] 98
4.16 Displacement, velocity and acceleration of the follower with a
      cycloidal cam   99
4.17 Picking stick bent by inertia force   102
4.18 Nominal and actual displacement velocity and acceleration of the
      shuttle   104

CHAPTER 5

5.1   View and diagrammatic representation of a four-bar linkage   107
5.2   Sley and crank beat-up mechanism, photograph and diagrammatic
      representation   108
5.3   Structures and linkage mechanisms   108
5.4   Geometry of the four-bar linkage   111
5.5   Equivalent four-bar linkage and 3 linkage-slider mechanisms   111
5.6   Displacement, velocity and acceleration of the simple harmonic
      motion   113

5.7 Calculation of 'dwell' clearance time on a loom — 113
5.8 Effect of crank-arm ratio on 'dwell' — 116
5.9 Path of bearded knitting needle in flat bed knitting machine — 117
5.10 Paths of various points on a four-bar linkage *[Reproduced from* Hrönes and Nelson *Analysis of Four-Bar Mechanism,* Wiley, 1951] — 118
5.11 Transfer mechanism *[Reproduced from* J. S. Beggs *Mechanism,* McGraw-Hill, 1955] — 118
5.12 Required needle movement on warp knitting machine — 119
5.13 Double beat-up mechanism — 120
5.14 FNF linkage mechanism for driving the needle bar — 121
5.15 Movement obtained by varying the parameters of the FNF linkage mechanism — 122
5.16 Eight-bar linkage for needle bar movement — 123
5.17 Choosing critical points on a required movement — 124
5.18 Ratchet and pawl — 124
5.19 Vernier ratchet and pawl — 125
5.20 Control by means of ratchet and pawl — 126
5.21 Cam and screw mechanism of flat bed knitting machine — 127
5.22 Required movement of the detaching rollers — 128
5.23 Mechanism for driving detaching rollers — 129

CHAPTER 6

6.1 Various forms of information storage — 132
6.2 Drum and information store on a half-hose machine — 134
6.3 Curved link chain selection on warp knitting machine — 135
6.4 Information store for Binary Cottons Patent machine *[Courtesy of* S. A. Monk Ltd] — 137
6.5a. Standard dobby with chain, and b. paper tape dobby *[Courtesy of* Geo. Hattersley Ltd] — 139
6.6a. Punched card label loom Jacquard, and b. photo-electric scanner for similar machine *[Courtesy of* Apparatus Factory Ltd] — 141
6.7 Grouping of needles for an 18-wale repeat — 142
6.8 Brinton drum selector — 143
6.9 Warp threading for repeating pattern — 143
6.10 Warp threading for non-repeating pattern — 144
6.11 Tappet selection on loom — 145
6.12 Single cylinder, single-lift Jacquard — 147
6.13 Movement of hook and knives in a single cylinder, single-lift Jacquard — 148
6.14 Movement of hooks and knives in a single cylinder Jacquard with movable bottom board — 149
6.15 Single cylinder, double-lift Jacquard — 150
6.16 Pulley method of averaging two knife movements — 150
6.17 Movement of hooks, knives and warp in open-shed Jacquard — 151
6.18 Hattersley V-dobby *[Courtesy of* Geo. Hattersley Ltd] — 152
6.19 Names of parts on V-dobby — 153

6.20 Movement of hooks, knives and heald shaft (warp) in V-dobby 154
6.21 Gear system in Knowles dobby
                [*Courtesy of* Crompton and Knowles Ltd] 155
6.22 Gear system in Leeming dobby 157
6.23 The Staubli V-dobby   [*Courtesy of* Staubli Bros. and Co.] 160
6.24 The paper tape selector mechanism on the Staubli dobby
                [*Courtesy of* Staubli Bros. and Co.] 161
6.25 Leverage system in a four-box mechanism 164
6.26 Six-box motion leverage system 165
6.27 Geneva mechanism 167
6.28 Locking action on Geneva mechanism 167
6.29 Indexer for paper dobbies 168
6.30 Geared indexer 169
6.31 Scroll indexer 170

CHAPTER 7

7.1 Diagrammatic representation of open- and closed-loop systems 174
7.2 Warp breakage detectors 177
7.3 Yarn breakage detector 178
7.4 Weft fork detector 179
7.5 Negative let-off tension control 180
7.6 a. disc, and b. gate yarn tensioners 181
7.7 Simple tension control 182
7.8 Weight and spring positioning of floating beam in let-off control 184
7.9 Hattersley light standard let-off motion 184
7.10 Shield and ratchet let-off mechanism 185
7.11 Principle of warp knitting machine positive let-off controller
    [*Reproduced from* W. F. Paling *Warp Knitting Technology*, Harlequin
      Press] 186
7.12 Details of warp knitting machine positive let-off controller 186
7.13 Raper autoleveller   [*Reproduced from the Textile Manufacturer*] 189
7.14 Autocount controller for woollen carding machines 190
7.15 Replacement of pirn selection mechanism 192
7.16 Diaphragm valve   [*Courtesy of* J. Blakeborough and Sons Ltd] 194
7.17 Programmed temperature controller
           [*Courtesy of* Mason-Neilan Regulator Company] 195

excessive wear. The two most important restraints are those which produce a turning or a sliding motion. A turning pair, or hinge, is one in which the two elements can move relative to each other about the axis of the hinge. A sliding pair is one in which one element slides on another; the elements comprising the pair may either be straight or curved. In Fig. 1.1 the picking stick and frame form a turning pair; all the other pairs shown are sliding pairs. It is easier to

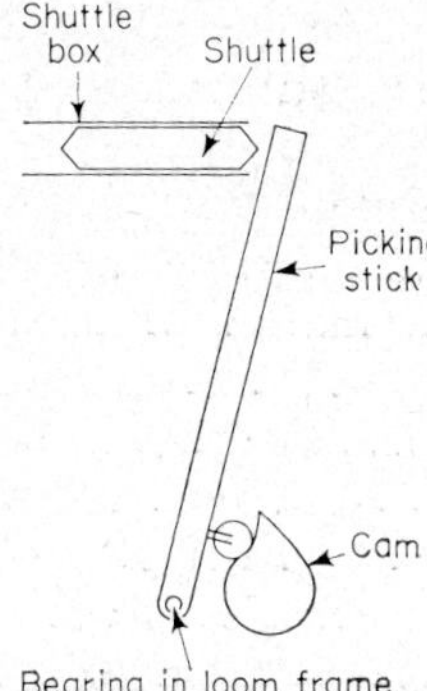

FIG. 1.1. Picking stick in loom, showing various parts.

construct a satisfactory turning pair than it is to make an efficient sliding pair, and to appreciate the reasons for this it is necessary to examine the nature of friction and wear, and the role lubrication plays in decreasing both friction and wear.

1.1 THE NATURE OF FRICTION

When two solid bodies are brought into intimate contact so that they touch, i.e. the atoms in the two parts approach each other to within distances of the same order of magnitude as the atomic spacing, the two bodies will, to all intents and purposes, become one, and any sliding between them will become impossible unless the bodies are broken along the direction of sliding. Under normal circumstances, two bodies do not touch in this sense except at very isolated points, due to the irregular nature of the surfaces when viewed on this atomic scale. The true area of contact between surfaces is thus very small. It has been shown that, for most metal bodies, the true area of contact between them varies directly with the force of the one body on the other and is to a first order of approximation independent of the irregularity of the two surfaces.

When bodies are rubbed against each other these small contact areas are formed and broken continually. The force needed to break the contacts results in the frictional forces normally found. The breaking results in the removal of some material from one of the surfaces, thus producing wear. It has been shown that even at quite moderate speeds this breaking of junctions results in relatively large quantities of heat being produced at the contact points, so that the surface temperature can readily exceed the melting point of the material of the surface. When this happens, catastrophic wear results. To prevent this friction and wear it is necessary to prevent the surfaces coming into contact. This can be done in three ways:

1 By interposing a very thin film—of approximately molecular thickness—between the surfaces. Such a film can only maintain its position between the two surfaces by molecular adhesion or actual chemical reaction between the material of the layer and of the surface.

2 By interposing a thick film of fluid under pressure between the surfaces.

3 By interposing a ball or roller between the two surfaces.

1.2 LUBRICATION BY MEANS OF CHEMICAL FILMS

In their natural state all metals have an oxide layer or a layer of some other contaminating substance on their surface. These naturally occurring films often provide considerable resistance to the adhesion of the surfaces. In fact, it is found that the coefficient of friction of degreased metal surfaces can be increased by a factor of about 10 when the oxide layer is removed by rubbing the surfaces in a vacuum. The wear resistance and coefficient of friction can be considerably lowered, however, by deliberately contaminating the surface with animal or vegetable oils. Their action depends on the presence in the oil of a chemically reactive compound. It has been shown that the most useful of these additives, which are the fatty acids, actually react with the surface to produce a metallic soap. While such a layer can produce large reductions in the coefficient of friction and wear rates, it must be remembered that the layer is continually being rubbed off, and there must therefore be available a reserve of oil to replenish the layer.

There are many applications, especially in textile machines, where it is difficult to maintain a continual supply of lubricant. For example, ring travellers, especially when used for cotton spinning, very quickly become covered with particles of fibres which adhere to the oil if the surfaces are lubricated in the normal way. One of the rubbing surfaces can be made from small particles of metal partially melted together. Such sintered metal contains many small holes which can be filled with oil, and during use this results in lubricant being supplied continually to the two surfaces by the rubbing action of one on the other. After several months' use the oil can be replaced by immersing the sintered material in oil. Such rubbing surfaces are also useful for mechanisms which are lightly loaded, and are therefore extensively used on looms in the measuring and selection mechanisms, where the relative movement is small.

Similar materials containing a solid lubricant, PTFE (polytetrafluorethylene), instead of oil in the pores are also used extensively. PTFE owes its low frictional properties to its low chemical reactivity, which results in very weak bonds being formed between the contacting surfaces. Similar reasoning applies to the use of white metal linings for bearing surfaces. The low strength of the white metal results in junctions which are broken easily.

The use of films or layers of substances between the surfaces is possible only when the loading is relatively light and the rubbing speeds not too high. Otherwise, the layer is broken and rapidly removed. Where large loads and high speeds are to be expected it is necessary to separate the surfaces by very much thicker films.

1.3 LUBRICATION BY THICK FILMS OF FLUID

Two surfaces can be separated by a fluid film if the pressure in the film is sufficient to maintain the load between them. This pressure can be built up in two ways. It can be produced by forcing the lubricant into the space between the two surfaces by means of an external pressure source, or it may be the result of the build-up of pressure due to hydrodynamic forces in the fluid present in the clearance space between the two rubbing surfaces. These two forms of lubrication are known as hydrostatic and hydrodynamic lubrication, respectively. The methods in which these principles are applied to the lubrication of mechanisms will be discussed in Sections 1.5 and 1.6 but the reason why pressure is built up by hydrodynamic forces will be considered now, as the principles involved are basic to all the applications discussed later.

Hydrodynamic lubrication is always the result of the pressure which is built up in a fluid forced to flow into a narrowing channel between two surfaces moving at different speeds. If we consider two surfaces at an angle to each other and submerged in oil, then it can be shown by means of the following argument that, if one surface is moved in a direction parallel to the surface of the second, a pressure gradient is built up between the thicker and thinner sections of the wedge-shaped film.

Fig. 1.2 shows two such plane surfaces submerged in oil. The upper one (which is inclined at an angle to the lower) is made to move in a direction parallel to the lower surface at a speed v.

Let us consider what happens to the fluid trapped between the two surfaces say at the section BB' shown in Fig. 1.2. The oil in contact with the upper moving surface at B will have a speed equal to that of the surface. On the other hand, the fluid in contact with the lower surface at B' will be at rest since the lower surface is stationary. Consequently, the fluid in any position D between B and B' must have a velocity intermediate between the speeds of the upper and lower surfaces. In other words, as D is moved from B' to B the velocity of the fluid will vary. This variation of velocity is called a velocity gradient across the section.

Suppose that at the section BB' the variation of the velocity across the section is linear so that the velocity gradient is constant as shown in Fig. 1.2. We can then find the shape of the velocity gradient at the section AA' as follows. Suppose the plates are of uniform width, w. The area of the section AA' (perpendicular to the plane of the paper) is $w.AA'$, and the total quantity of fluid passing through this section in unit time is:

$$\bar{v}_a.w.AA'$$

where $\bar{v}_a$ is the average fluid velocity over the section AA'. In the same way, the total quantity of fluid passing through the section BB' is:

$$\bar{v}_b.w.BB'$$

But these two quantities must be equal, for otherwise there would be a build-up of fluid in the wedge formed by the two surfaces. Thus we must have:

$$\bar{v}_a . w . AA' = \bar{v}_b . w . BB'$$

or

$$\frac{\bar{v}_b}{\bar{v}_a} = \frac{AA'}{BB'}$$

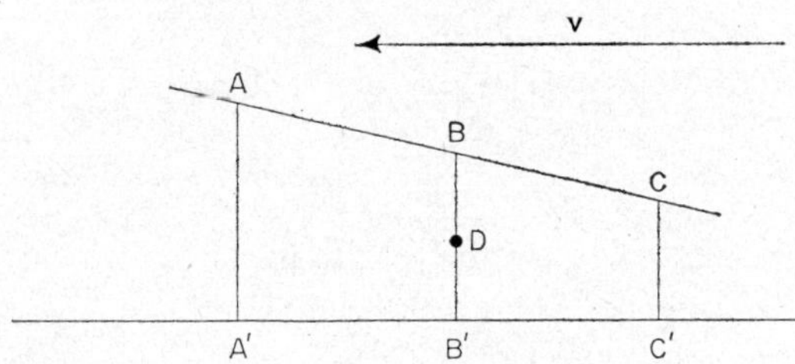

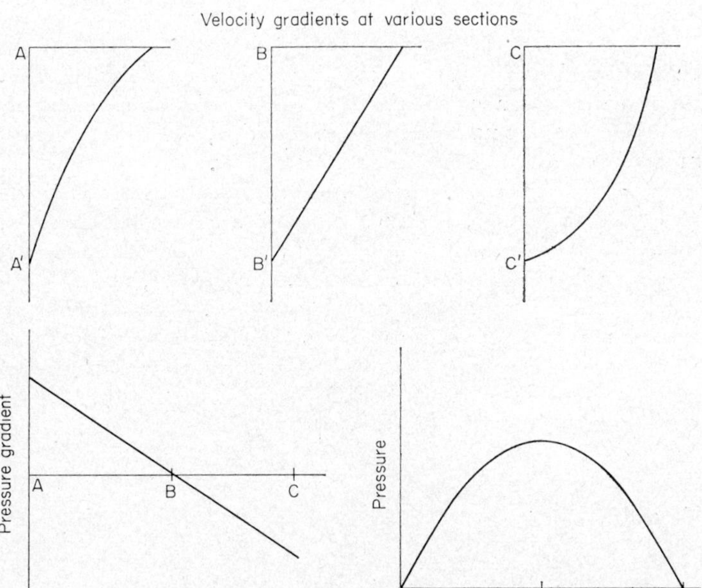

FIG. 1.2. Flow of lubricant between two inclined plane surfaces.

Since AA' is greater than BB', $\bar{v}_b$ must be greater than $\bar{v}_a$, i.e. the average velocity at BB' is greater than that at AA'. Now since the fluid velocities at A, B and at A', B' are the same (being equal in both cases to the velocities of the surfaces) it follows that the velocity curve at AA' is concave, as shown in Fig. 1.2. In the same way, it can be shown that the velocity curve at BB' is convex.

The basic equations for the behaviour of a viscous fluid can be manipulated into a form which shows that the pressure gradient *along* the film is directly proportional to the product of the viscosity of the lubricant and the rate of change of velocity gradient *across* the film. This rate of change of velocity gradient is constant across any section but varies from section to section. At BB' this rate of change of gradient is clearly zero, while at AA' it is positive, and at CC' negative. The pressure gradient at each point must therefore be similar to that shown in Fig. 1.2, and clearly, as the pressure at the two ends of the slider must be zero, the pressure in the space between the two surfaces must be of the form shown in Fig. 1.2. This pressure can clearly support a load placed on the moving surface. The film is therefore able to keep the two surfaces apart even under load. From the analysis outlined above, it follows that no load can be carried in this way if the two surfaces are parallel to each other. If they are parallel, the velocity gradients at all points will be the same as at BB', and hence the pressure gradients and pressures are zero at all points.

1.4 THE USE OF ROLLING INSTEAD OF SLIDING CONTACT

By the use of rolling instead of sliding contact between the elements of a machine, it is possible to maintain a low frictional resistance with a minimum of lubrication. This arises from the nature of rolling. Friction, as normally defined, arises only from the sliding of two surfaces past each other. In a rolling contact, however, the two surfaces are moving apart at right-angles to the direction of relative movement of the two bodies, and no sliding takes place. It therefore appears, at first sight, that no frictional resistance or wear should take place at a rolling contact, and hence no lubrication is required. In practice, however, this is not so since some sliding occurs even at a rolling contact. This arises from the distortion of the surfaces where the ball meets the contacting surface under load. As can be seen from Fig. 1.3, at the point of contact the two surfaces penetrate

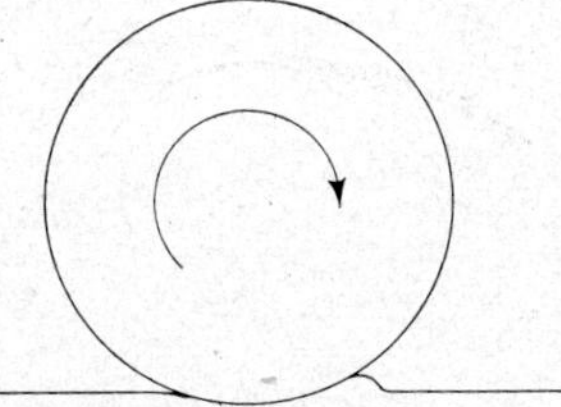

FIG. 1.3. Contact between rolling body and a plane.

into each other and, when rolling takes place, a wedge of metal in front of the rolling body has to be continuously moved with it. In so doing a small amount of sliding takes place. Because of this sliding, it is necessary to provide some form of surface film lubrication. Furthermore, lubrication also helps to exclude dirt and metal chips from the contact region; the surfaces would be destroyed rapidly by the abrasive action of such contaminants.

Because of the small-scale slipping and the presence of the lubricant, it is found that the 'frictional' resistance at a rolling contact is not zero, firstly, because of the frictional resistance to this small-scale slipping, and, secondly, because the lubricant is being continually forced out of the space in front of the rolling member, a viscous resistance to movement results. It has also been shown that the continuous deformation of the two surfaces results in some loss of energy because of the hysteresis loss when metal is compressed and released. All these losses, however, are relatively small in a well-maintained rolling pair, and co-efficients of friction for such contacts usually lie in the range from 0·001 to 0·005.

It has already been noted that any lubricant present on the rolling surfaces will tend to be 'squeezed' out by the rolling contact. This action will produce a hydrodynamic pressure in the oil, but in the past it has been considered that this pressure was inadequate to separate the two surfaces. Two factors, however, make such a separation possible under certain circumstances. Firstly, the fact that the viscosity of most lubricants rises very steeply with increase in pressure; for example, the viscosity of a typical mineral oil will rise one hundredfold when the pressure rises from atmospheric pressure to 16 tons/in². Secondly, the distortion in the shape of the rolling surfaces produces a shaped 'channel' which will enhance the pressure produced during squeezing out of the lubricant. In fact, it is found that many rolling contacts, such as occur in ball bearings, gears and roller cams (see later), are separated by a very thin, very viscous film of lubricant. This form of lubrication is known as elasto-hydrodynamic lubrication, and for further details of this very interesting but very complex form of lubrication the reader is referred to the bibliography at the end of this chapter.

1.5 THE DESIGN OF TURNING PAIRS—JOURNAL AND ROLLER BEARINGS

To produce a turning pair it is necessary to restrict the movement of one of the elements so that it can only rotate with respect to the second element of the pair. The standard method of doing this is to attach a cylinder to one of the elements, known as the shaft, and pass it through a hole drilled in the second element, known as the bearing. The hole or bearing diameter is always greater than the shaft diameter, and if the clearance space is small and filled with oil or grease, the bearing is known as a journal bearing. If the clearance space is large and is filled with balls or rollers to produce a rolling contact between the two moving surfaces, the bearing is known as a ball or roller bearing. It should be noted that, in general, bearings allow the shaft to move in two ways, in other words, it has two degrees of freedom. The shaft can rotate in the bearing or it can move axially in and out of the bearing. This second movement can be prevented by mounting the shaft between two collars as shown in Fig. 1.4. This method of preventing lateral movement can only be used where the force tending to move the shaft axially in the bearing is small. Other methods of withstanding axial loads will be discussed later, but let us for the moment consider the problem of lubricating the curved inner surface of the bearing, and the outside of the shaft.

It is clear that these can be lubricated by the use of chemical films or by the use of sintered metal containing oil or PTFE for the bearing surface material.

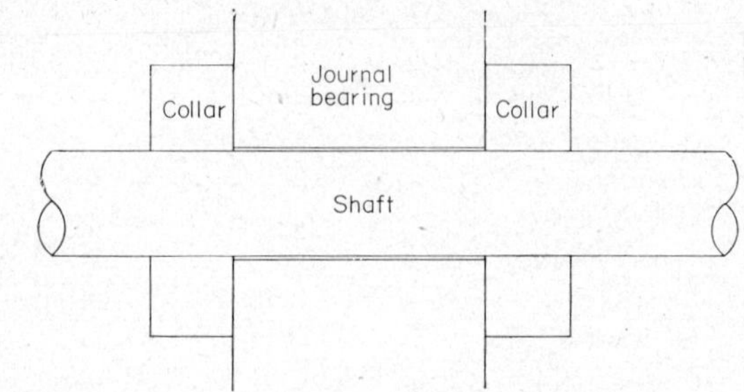

FIG. 1.4. Restraints on a shaft.

As has previously been mentioned, such bearings are frequently used for lightly loaded mechanisms which are relatively inaccessible and cannot easily be supplied with a continuous flow of lubricant. Journal bearings, however, are particularly suitable for forming hydrodynamic or hydrostatic pressure films. This arises directly from the geometrical relationship between the shaft and the bearing. Let us first of all consider the hydrodynamic lubrication of journal bearings. The clearance space between shaft and bearing is kept filled with oil. When the shaft is stationary and under load, it will lie at the bottom of the bearing as shown in Fig. 1.5a, and the clearance space which is filled with oil has the sickle shape shown in this diagram. It can be seen that this clearance consists of a wedge of oil. As soon as the shaft is rotated, a build-up of pressure will occur in these oil wedges so that the shaft is lifted off the bearing surface. If the load is relatively high and the speed not too great, the shaft will remain in a position such as that shown in Fig. 1.5b where the shaft is just clear of the bearing surface. If the load is very small or the speed very high, only a very slight wedge of oil is sufficient to support the load, and the shaft therefore moves towards, but never reaches, the centre of the bearing, as shown in Fig. 1.5c.

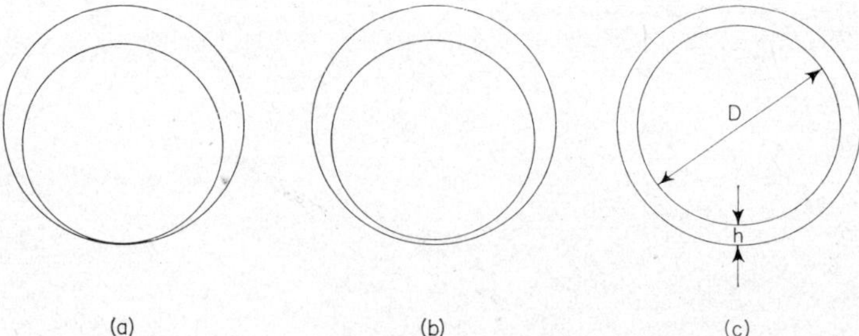

FIG. 1.5. Position of shaft in journal bearing under differing loads and at differing speeds.

It can be shown from this type of argument that in a journal bearing, the load that can be supported by the fluid film depends on the speed of rotation of the shaft, the viscosity of the lubricant and the relative clearance of the shaft in the bearing (h/D), see Fig. 1.5c. It follows, therefore, that the relative clearance is dependent on the load per unit area of the bearing, W, the speed of the shaft, N, and the viscosity of the lubricant, Z, which has the dimensions of force $\times$ time/(length)^2.

In considering the behaviour of journal bearings it is also necessary to know the power loss in such a bearing. The power loss is due to the viscous resistance of the lubricant, and depends on N, Z and W. This power loss can also be expressed in terms of the coefficient of friction of the bearing. The coefficient

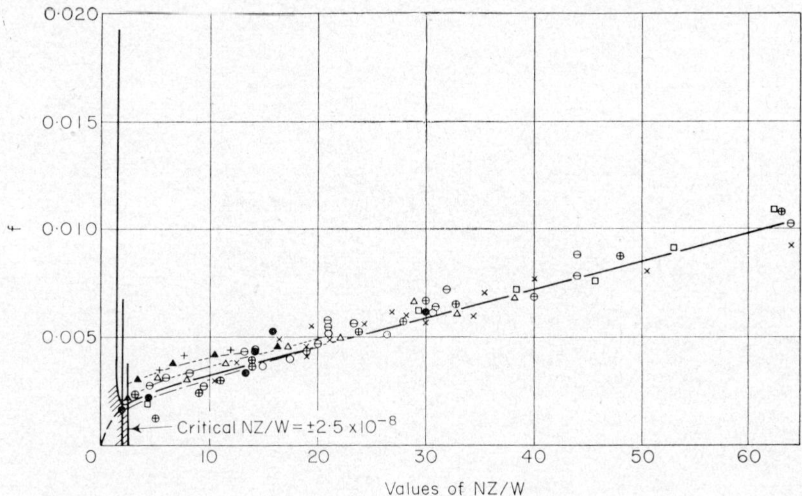

Pressures in kg. per sq. cm.: $+ = 58$, $\blacktriangle = 46$, $\circ = 38$, $\triangle = 31$, $\times = 14$, $\bullet = 7$, $\ominus\oplus = 7 < p < 38$

FIG. 1.6. Experimental results for friction in a journal bearing, obtained by Prof. A. Tenot (by courtesy of Inst. of Mechl. Eng.).

(Reproduced from *Proceedings of Inst. of Mechl. Engineers: General Discussion on Lubrication*, Vol. 1, p. 321, 1936.)

of friction, f, of the bearing has no dimensions, and the reader can soon check that the ratio NZ/W is also dimensionless. It follows that f must be a function of NZ/W only. In other words, if we plot f against NZ/W a single graph will result which is independent of the particular values of N, Z or W chosen, but depends rather on their combination in the group NZ/W. (This group is known as the Sommerfeld number.) Fig. 1.6 shows a set of experimental results plotted in this way, which confirms that this approach is valid over the range of values chosen. It can be shown, in fact, that the coefficient of friction is directly proportional to $\sqrt{NZ/W}$.

It is found, however, that at low values of NZ/W this relationship ceases to apply, the type of curve found being shown in Fig. 1.7. To appreciate the reason for this change in behaviour it is necessary to note that the clearance, h, is large when the shaft moves towards the centre of the bearing—this, as we have seen, occurs at low loads and high speeds; in other words, h is large when NZ/W is large, and small when NZ/W is small. There is, in practice, a minimum value of the clearance which is dictated by the smoothness of the bearing and shaft surfaces. If h is too small the film will be bridged in places by the irregularities of the surfaces, and metal to metal contact will take place. This results in a large increase in frictional drag and accounts for the large rise in friction at low values of NZ/W, as shown in Fig. 1.7. The point of minimum friction, point A on Fig. 1.7, represents approximately the beginning of the onset of metal to metal contact, which must be avoided if wear is to be minimised. It can be seen, therefore, that a choice has to be made between high friction at large values of NZ/W and the danger of high wear at low values of NZ/W.

The choice of a suitable value for NZ/W depends upon the geometry of the bearing, the type of load on the bearing (whether it is constant or rapidly varying), and on the surface finish of the surfaces. This makes it difficult to give a

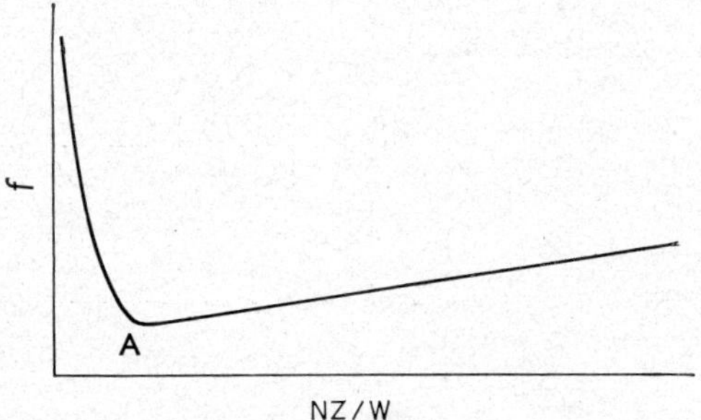

FIG. 1.7. f versus NZ/W.

fixed rule for the correct choice of a lubricant, but it follows that for high loads and low speeds a high-viscosity oil or grease will be required, while in a high-speed system a low-viscosity oil will be needed.

All hydrodynamically lubricated bearings suffer from the disadvantage that during starting up contact takes place between the two surfaces because the value of NZ/W is small. This is particularly true of certain applications where, either because the speed of operation is very high or because the friction must be kept very low, it is desirable to use a very low-viscosity lubricant (usually air since it has the lowest viscosity of all readily available fluids). However, lubricants with very low viscosities do not usually provide any lubrication by the formation of chemical films and there is therefore a tendency for the bearing to seize, or become welded to the shaft, during the start-up period. This difficulty does not, however, occur with hydrostatic lubrication.

To achieve hydrostatic lubrication it is necessary to supply the lubricant to the bearing in such a way that the shaft will tend to float in the centre of the clearance space. Fig. 1.8 shows the principle on which most hydrostatic systems operate. Oil, or any other suitable fluid, is supplied under pressure, usually of the order of 100 lb/in², into the passage A and escapes from there through the three orifices marked O into the clearance space between the shaft and bearing. The fluid can then escape from the pockets marked P only by passing along the shaft, through the very small clearance space, and escaping to air. If the gap between the shaft and the bearing becomes large near one of the orifices, the pressure on this side of the bearing will drop and there will be a resultant thrust from the other two orifices where the gap has decreased and the pressure therefore increased. As a result, the shaft will be thrust back towards the centre, and will tend to float equidistant from all three orifices.

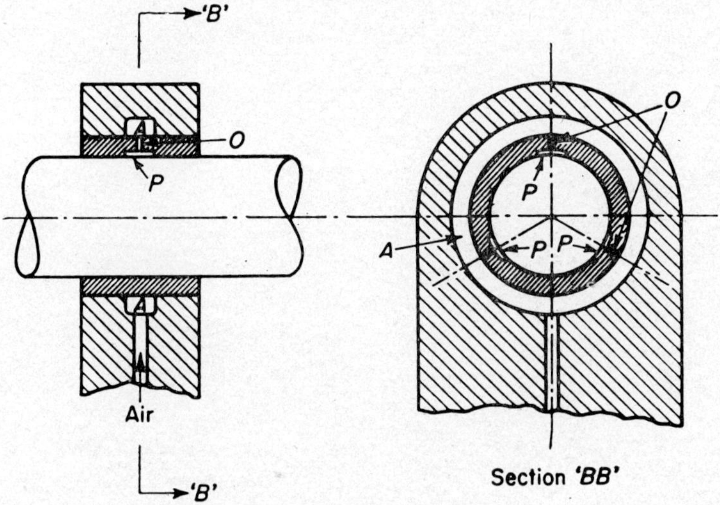

FIG. 1.8. Hydrostatic Bearing.

(Reproduced from Morrison and Crossland, *An Introduction to the Mechanics of Machines*, p. 445, Longmans, 1964.)

While such bearings are simple in principle, there are many practical difficulties in their construction. For example, if air is to be used as the lubricant, the space between shaft and bearing has to be very small; this requires accurate machining and a very high-quality finish on the bearing and shaft. In addition the design of the passages from the orifices to the clearance space is difficult, and is more of an art than a science. Despite these difficulties, such bearings have received considerable attention because of their unique properties. When used with air they provide an almost frictionless bearing. It should be noted also that this very low friction occurs even when the shaft is stationary, in contrast to the hydrodynamic bearing which has a low frictional restraint only when running at its designed speed. The second advantage of these bearings is that they do not wear during start-up and shut-down of the machine.

Because of its low frictional properties the hydrostatic bearing is used extensively for instruments, and it has been proposed for use in supporting ring spindles that operate at high speeds. The relatively high cost of such bearings has, however, prevented any more extensive use at present. A further use for such bearings in textile machinery is as bearings for high-speed false twister frames. The false twister tube operates at 500,000 r.p.m. and normal bearings are quite unusable at this speed.

A more common bearing for this purpose, however, has been the so-called magnetic bearing. This is not, in fact, a bearing in the normal sense although it is a method of restraining a cylinder so that it rotates about an axis. Fig. 1.9 shows the positioning of the false twister tube. It is driven by means of the two rubber discs, *A* and *B*, and it can be seen that if the tube can be kept in contact with the discs at all times the tube must rotate as if in a bearing. The tube is kept

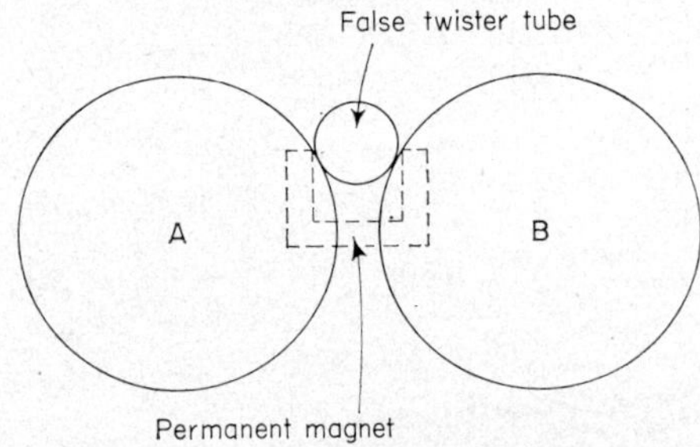

FIG. 1.9. Magnetic Bearing.

in contact with the rubber discs by means of a magnetic force provided by a permanent magnet. By this means a restraint has been provided without physical contact on the tube.

The majority of the methods so far considered have depended for their action on the presence of some lubricant. In addition, if large loads are carried it is found that the journal bearing must be fairly bulky, and since it must be supplied continuously with oil some means must be available for removing the excess oil from the bearing. There are many textile applications where it would theoretically be possible to use such bearings, but the cost would be very high, and they are consequently impractical. A partial solution, such as using a grease-filled bearing or some sintered bearing material, is sometimes possible, but where it is necessary to maintain accurate location of the shaft over long periods, as in the top rollers of spinning frames, or when large loads are to be expected, it becomes necessary to separate the surfaces by means of ball or roller bearings.

There are many different forms of rolling contact bearings suitable for different applications, but they all fall into a few distinct classes. The types of bearing are divided into two, depending upon whether cylinders or spheres are used as the intermediate element. Cylindrical, or roller bearings as they are called, are generally used for heavier loads. In addition, such bearings are further subdivided depending on whether the bearing can position the shaft

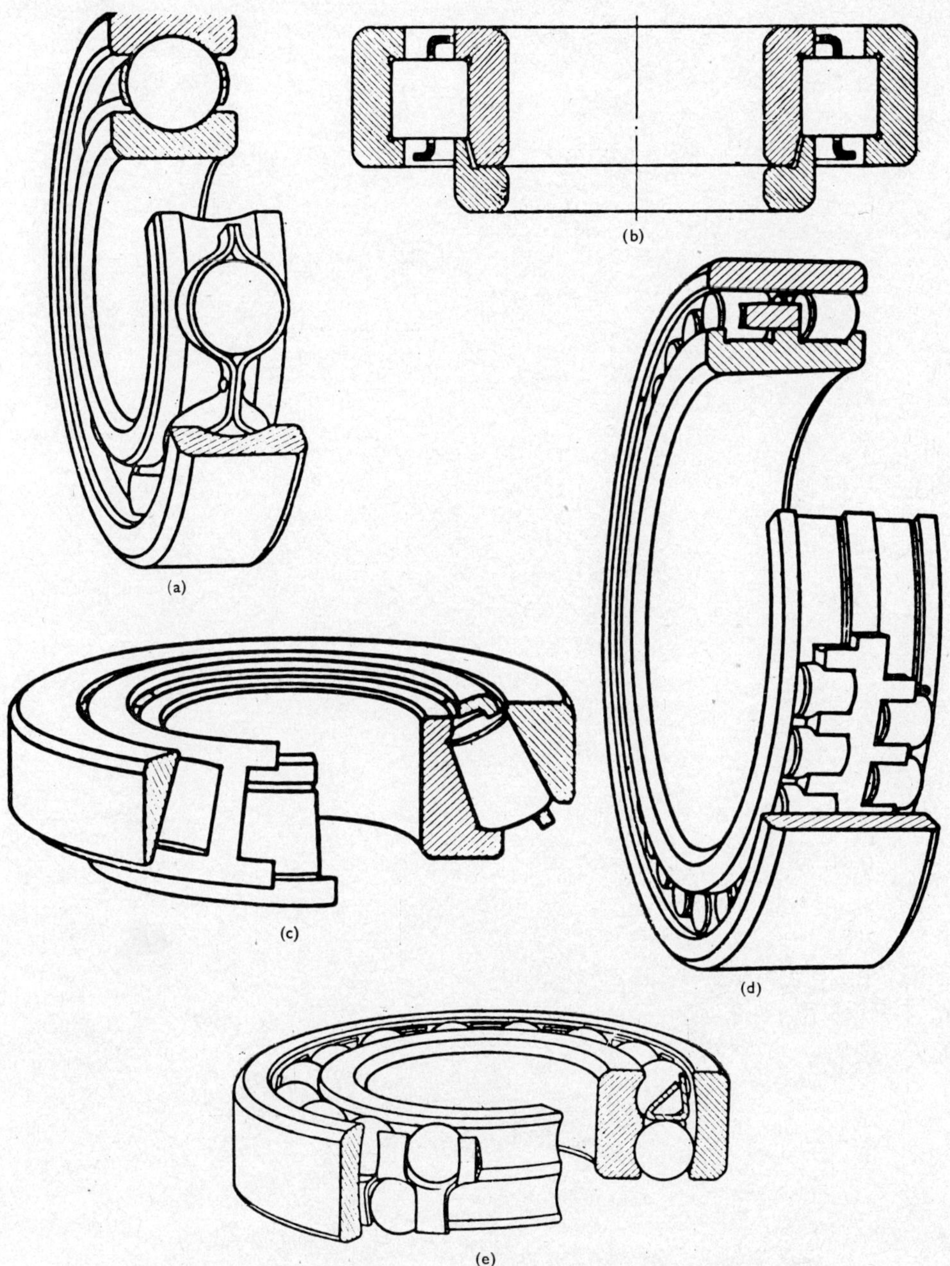

Fig. 1.10a. Deep groove ball; b. roller; c. thrust; d. cylinder, and e. self-aligning ball bearing (by courtesy of S.K.F. Ltd).

against forces along the shaft, and whether they can support such forces in one or both directions. In this respect they differ from journal bearings where, as we have noticed, a separate means must be provided for preventing the shaft from moving along its axis in and out of the bearing and, in so doing, support any load along the axis, or thrust, as it is called.

Figs. 1.10a and b show the standard form of deep groove ball and roller bearings. Because the rolling element in these bearings lies in two deep grooves, one in the part of the bearing attached to the shaft, and one attached to the stationary section, these bearings will support both axial and radial loads. It might be noted also that the load is always carried by the ball by means of rolling contact with the groove. The cylindrical bearing, however, will only take radial loads in rolling contact, axial loads will be supported between the flat sides of the cylinders and the groove where the contact is sliding. This roller bearing is therefore not as suitable as the ball bearing for taking axial loads, and the tapered roller bearing shown in Fig. 1.10c is specially designed to take axial loads.

Fig. 1.10d shows a cylinder bearing which allows the shaft to move axially. It is used when it is desirable for the shaft to be able to move axially, for example where it is necessary to remove the shaft for maintenance purposes—as in a spindle bearing.

Both journal and roller bearings are rarely used singly to produce a turning pair because of the effect of wear on the positioning of the shaft axis. The shaft axis can move slightly, even in a new bearing, because some clearance must be

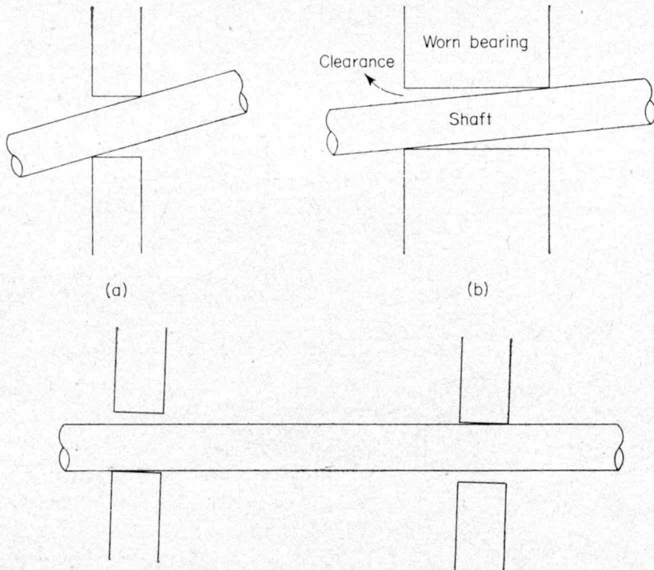

FIG. 1.11. Effect of wear on: a. a short bearing, b. a long bearing, and c. a two bearing support.

allowed between the moving parts. As the bearing is used, however, both the grooves and the curved bearing surfaces become worn and the possible movement of the shaft axis becomes greater. Fig. 1.11b shows the effect of wear on the possible inclination of the shaft axis to the bearing axis, and it is clear from Fig. 1.11a that for the same clearance the angular movement of the shaft will be greater if the axial bearing length is shorter. As journal bearings are usually much longer than roller bearings this problem is clearly more acute with roller bearings. This problem is usually overcome by mounting the shaft between two roller or journal bearings some distance apart along the shaft, as shown in Fig. 1.11c. Such positioning of the shaft must be carried out with care. The shaft passing through the first bearing is already constrained in its movement and the second bearing, which is providing some spare restraint to cope with the effects of wear, must be correctly positioned if the shaft is not to be bent, so producing very large forces on the bearing itself. This type of restraint is known in general as over-restraint and, while necessary for practical purposes, requires forethought in its application.

As an example, let us consider a shaft carrying a load between two such bearings. As a result of the load the shaft must be deflected, but due to the

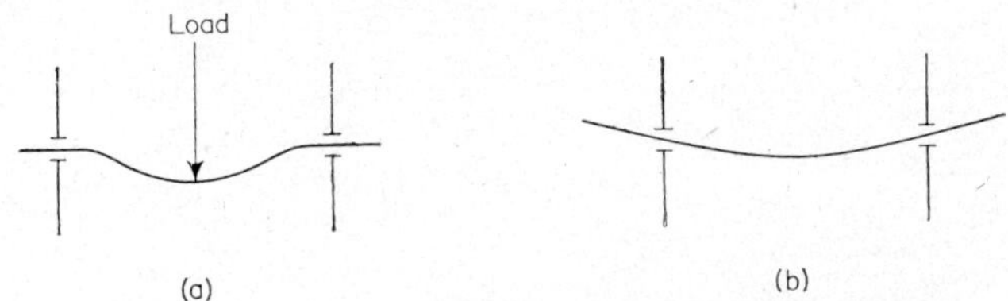

FIG. 1.12. Position of shaft under load with fixed and
self-aligning bearings.

restraint of the bearings, this deflection has to be zero at the bearings and, what is more, the direction of the shaft at the bearings must remain the same as it was before deflection. The deflected shape, therefore, is similar to the curve shown in Fig. 1.12a. The turning forces on the bearings under such conditions are considerable and unnecessary, since it is usually not necessary that the shaft should be positioned in this way. Under such circumstances, self-aligning roller bearings such as that shown in Fig. 1.10e can be used. These balls do not run in a deep groove but against a shallow curve, the centre of which is at the centre of the shaft axis. As a result, the shaft inner part of the race and the balls can rotate about the centre of the bearing. If two such bearings are used to mount a shaft carrying a centre load as shown in Fig. 1.12a the shaft will deflect as shown in Fig. 1.12b, thus greatly reducing the load on the bearings.

The type of bearing, whether ball, roller or journal, will therefore be determined by the loading on the shaft, the positioning requirements of the shaft and the need for servicing, etc. Fig. 1.13 shows, as an example of a complete bearing unit, the means used to support a spindle for a ring spinning frame. Because the spindle rotates at high speed it must be capable of flexing under load

due to the unbalanced nature of the package, and must also be capable of being
easily removed. The upper bearing is a cylindrical roller bearing which carries
the major part of the load, while the lower bearing is a hydrodynamically lubricated conical bearing which provides the extra rigidity, and carries the vertical
load due to the weight of the spindle. The inner casing which carries these two

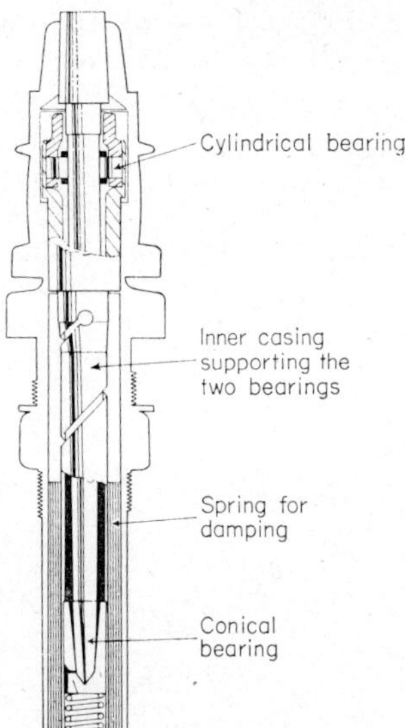

FIG. 1.13. Ring spinning bearing
unit (by courtesy of S.K.F.
Ltd).

bearings is flexible and allows the shaft to bend between the two bearings under
load. This bending, however, must be damped out to prevent high-speed vibration, and the inner casing is therefore surrounded by a spring; the whole bearing
being completely immersed in oil. When the shaft bends the movement of the
spring through the oil dampens out any vibration. This example has been
included to show that while the basic bearings and their action may be simple,
their application may require considerable skill and ingenuity.

1.6 THE DESIGN OF SLIDING PAIRS

Sliding pairs can be designed so that one element moves along the other in a
straight line or along a curved path. An example of the first kind is the shuttle
moving in the box while one of the second kind is the picking stick moving round
the cam (see Fig. 1.1). In all cases, however, it is considerably more difficult
to obtain good separation of the surfaces to reduce wear and friction for sliding
pairs than for turning pairs. Thus Fig. 1.14 shows a means of obtaining a
linear sliding pair which makes use of two journal bearings. While these bearings
can be supplied with oil they will not produce a hydrodynamic film since the
sliding of a shaft in a journal bearing will not produce a wedge-shaped oil film.

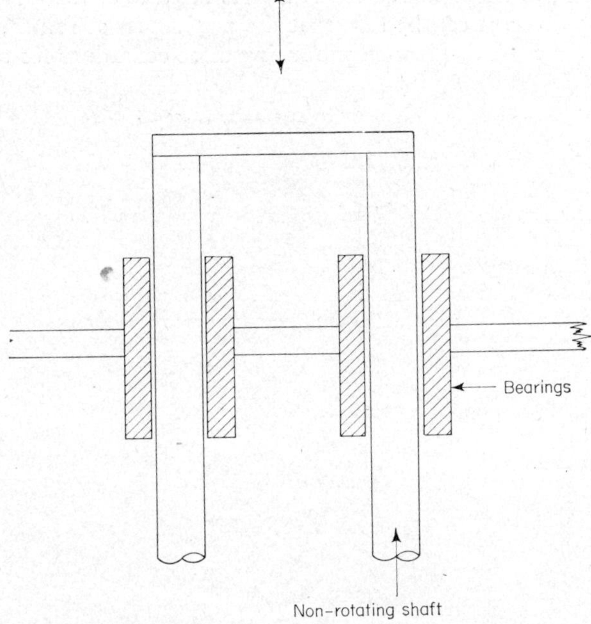

FIG. 1.14. Sliding pair using journal bearings.

Hydrostatic lubrication of sliding pairs is possible but rarely used because of cost. The most common method of separating the surfaces is to interpose a roller. Thus the cam in Fig. 1.1 will usually press against a roller attached to the picker stick. Further examples of such use of rollers in cam systems are given in a later chapter on cams. To obtain a linear motion a version of the ball bearing shown in Fig. 1.15 can be used. The system shown in Fig. 1.14 could

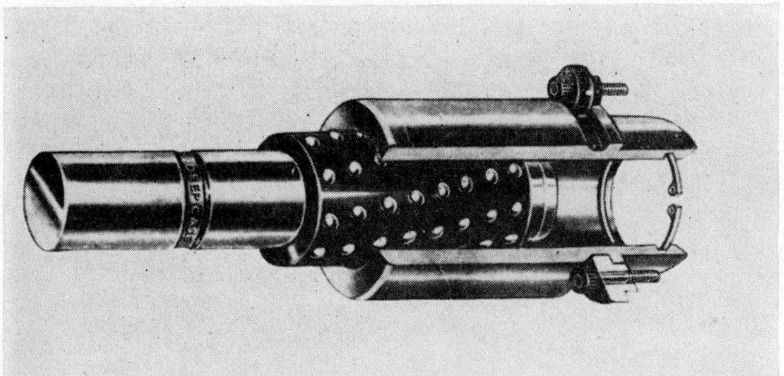

FIG. 1.15. Linear ball bearing (by courtesy of Rotolin Bearings Ltd).

C

also be modified to take sintered metal inserts in place of the journal bearings. In general, however, it can be said that the production of robust, trouble-free sliding pairs is more difficult than producing these conditions for turning pairs.

Suggested Further Reading

F. P. BOWDEN and D. TABOR. *The Friction and Lubrication of Solids.* Oxford, Clarendon Press, 1954.

F. T. BARWELL. *Lubrication of Bearings.* Butterworth, 1956.

Symposium on Lubrication of Textile Machinery, 1967, *Proc. of Inst. of Mechl. Eng.*, vol. 181, part 3L.

D. DOWSON and G. R. HIGGINSON. *J. Mech. Eng. Sci.*, vol. 1, no. 1, vol. 2, no. 2 and vol. 4, no. 2. (Papers on Elasto-hydrodynamic Lubrication.)

CHAPTER 2

THE TRANSMISSION AND MODIFICATION OF ROTATIONAL MOTION—FRICTIONAL DRIVES

2.0 INTRODUCTION

THE prime mover in any textile machine is almost invariably an electric motor which will rotate a shaft at some relatively fixed speed. This rotational motion must be transmitted to a large number of shafts, situated in various parts of the machine, each of which probably has to be driven at a different rotational speed. In the first chapter on this topic it is proposed to consider only those mechanisms that transmit rotational motion by the use of friction. In Chapter 3 more positive methods of transmitting rotational motion will be considered. Our main concern in the present chapter is to consider the transmission of rotational motion from one parallel shaft to another while at the same time altering its magnitude.

2.1 DEFINITION OF ROTATIONAL SPEED

It is common practice to define the rotational speed of a shaft in terms of the number of revolutions made by the shaft in unit time, the most common units used being revolutions per minute. This method, while fairly straightforward, is not the standard method used in the scientific literature.

The standard description of rotational speed is in terms of the angle turned by the shaft in unit time, the angle being expressed in radians, not degrees. Since the shaft rotates through 2π radians (or $360°$) in one revolution, it is evident that the angular speed, ω, is equal to 2π times the revolutions per unit time. It is usual to express the angular speed, ω, in units of radians per second; therefore if N is the number of revolutions per minute, it follows that:

$$\omega = \frac{2\pi N}{60}$$

(Note: ω is approximately one tenth of N.)

2.2 FRICTIONAL DRIVES BETWEEN TWO PARALLEL SHAFTS

The simplest method of transmitting rotational motion from one shaft to another that is parallel to the first is to fix two cylinders or discs on the two shafts such that the two discs touch (see Fig. 2.1). If the discs roll on each other and do not slip where they touch, the surface speed of the two discs must be the same. If the r.p.m. of disc 1 (Fig. 2.1) is N_1 and its radius r_1 then its surface speed will

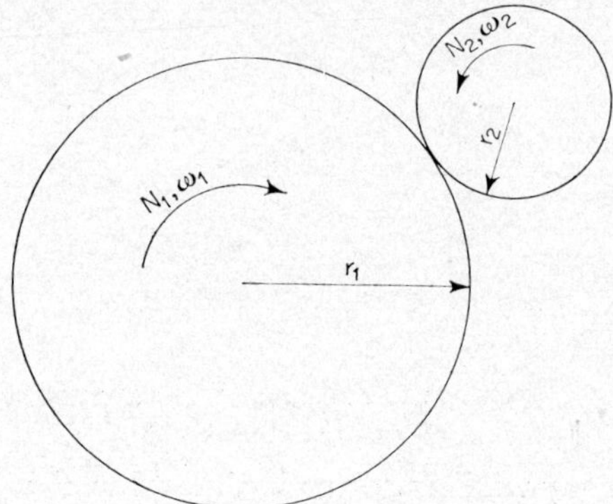

FIG. 2.1. The transmission of rotational motion between two discs.

be $N_1 \times (2\pi r_1)$ or $2\pi r_1 N_1$ in units of distance per minute. Alternatively, if the angular speed is ω_1 then the surface speed is $\omega_1 r_1$ in units of distance per second. Since the surface speed of both cylinders is the same, we must have:

$$2\pi r_1 N_1 = 2\pi r_2 N_2$$

or

$$\omega_1 r_1 = \omega_2 r_2$$

Hence:

$$\frac{N_1}{N_2} = \frac{\omega_1}{\omega_2} = \frac{r_2}{r_1} \qquad (2.1)$$

This equation shows that the rotational speeds of the two rollers are inversely proportional to the radii of the two cylinders. Equation (2.1) is of considerable importance, as it will be found to apply, in one form or another, to all the methods available for transmitting rotational motion between parallel shafts.

It should be noted that as a result of the rolling action, shaft 2 rotates in the opposite direction to shaft 1 so that both the speed and its direction have been altered.

It can also be seen that $r_1 + r_2$ is the distance between the centres of the two shafts, and this fact allows us to calculate the size of the wheels needed to obtain a given speed ratio. For example, if we have two shafts 4 in apart, and wish to drive them at a speed ratio of 6:1 then:

$$r_1 + r_2 = 4$$

while from Equation (2.1):

$$\frac{\omega_1}{\omega_2} = 6 = \frac{r_2}{r_1}$$

Hence:

$$6r_2 + r_2 = 4$$

or

$$r_2 = \tfrac{4}{7}''$$

therefore:

$$r_1 = 3\tfrac{3}{7}''$$

This form of drive is rarely used outside the textile industry because, in general, it is not capable of transmitting large horsepowers. It does, however, have several applications in the textile industry. There are several instances, for example, where the second cylinder might vary in radius, usually because yarn or fabric is being wound onto it.

It is essential in these cases to maintain a constant surface speed on this cylinder, and not a constant rotational speed. To take a specific example, Fig. 2.2 shows a cone being wound on a rotary traverse winder. The cone is

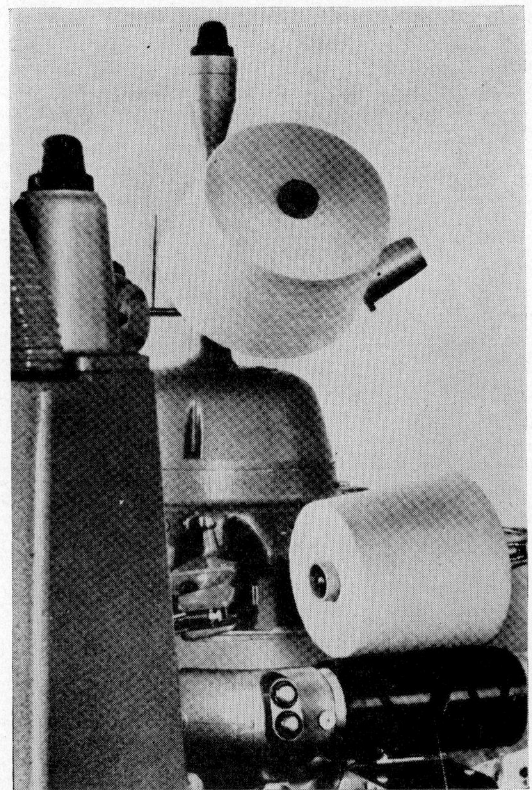

FIG. 2.2. Cone winder (by courtesy of Schweiter Engineering Works Ltd).

being directly driven by friction against the winding drum. This ensures a constant surface speed of the cone and consequently a constant speed of yarn winding onto the cheese. If the shaft that supports the cheese had been driven by some other means at constant speed, the surface speed of the cone would have increased rapidly as the radius of the cheese increased. It is, of course, possible to provide some form of varying speed drive to the shaft supporting the cheese which could compensate for any change in cheese diameter. Such mechanisms are available, but the simple solution of a direct surface drive is always preferred when it is practicable.

Similar reasoning applies to the use of direct friction drives for the top rollers of mangles, drafting rollers, etc. In all these applications it is necessary to obtain a grip between two pairs of rollers for gripping or squeezing fibres or cloth. To obtain such a grip it is necessary to distribute the load between the rollers over a reasonable area. This is usually accomplished by making the lower cylinder or roller of steel or some other hard material, while the upper roller consists of a layer of soft rubber, plastic, or cork material on a hard base. The diameter of the soft roller at the point where the two rollers meet varies with the

FIG. 2.3. Loading system for drafting rollers (by courtesy of Prince-Smith and Stells Ltd).

pressure applied, and if the surface speed of the two rollers is to remain the same at all times, it would be necessary to drive the top roller at varying speeds depending on the pressure applied. This is needlessly complex, and such roller pairs are always operated by driving the lower, hard surface roller at a constant speed, and allowing the upper roller to be driven by friction from the lower roller.

To operate such a drive successfully it is necessary to apply an even load to the top roller. Fig. 2.3 shows a simple loading system used in worsted spinning. The load is applied centrally between two pairs of rollers through a spring. By screwing the spring down to various heights it is possible to obtain varying loads on the rollers. As the load is applied centrally it is much easier to ensure that the load is evenly distributed across the face of the drafting rollers.

Where only one pair of rollers is involved, as in a mangle, a much more complex solution is necessary and Fig. 2.4 shows the principle of the Farmer Norton level pressure mangle. In this loading system it is necessary not only to apply similar loads to the two ends of the roller but it is also essential to prevent or compensate for the effect that such loading normally has on the shape of the rollers. The usual effect of loading at the ends would be to bend the top roller, with the result that the pressure between the rollers varies along their length; it is greatest at the ends of the rollers and is a minimum at their centre. The method chosen by Sir James Farmer Norton and Co. Ltd. to overcome this problem is to apply the load in such a way that the bending of the top roller is minimised. It

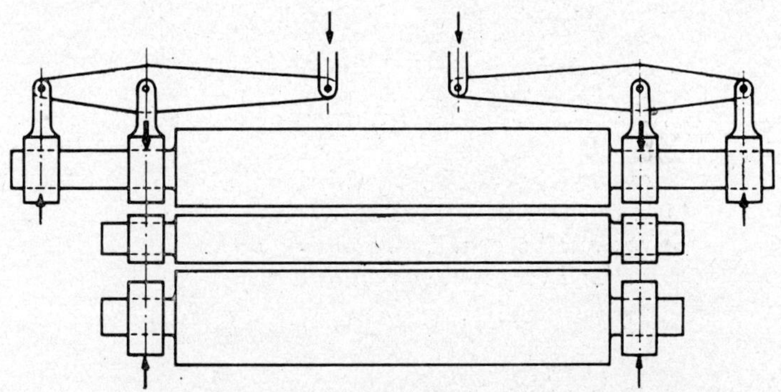

FIG. 2.4. Loading system on pressure mangle (by courtesy of
Sir James Farmer Norton & Co. Ltd).

can be seen that the behaviour of such frictional drives is of considerable importance in textile machinery, and for further details the reader is referred to the separate reading list on mangles at the end of this chapter.

There is a further use for friction drives which has already been mentioned. Fig. 1.9 shows the drive to a false twister tube. This tube rotates at 500,000 r.p.m., and it is found that the friction drive is the most suitable means of obtaining such high speeds because of its smooth action. All other forms of drive are discontinuous in one way or another, thus producing vibrations in the system which are excessive at such very high speeds.

2.3 BELT DRIVES

One of the major disadvantages of the direct friction drive is that it can only be applied to shafts which are reasonably close together. A friction drive can, however, still be used for shafts which are far apart, by using cylinders or pulleys on the two shafts which do not touch, but are connected by means of a belt. Fig. 2.5 shows the general principles of a belt drive. If the belt does not slip on the pulleys and the belt is inextensible, the surface speed of both rollers is the same as the surface speed of the belt, and hence the same conditions apply as for Equation (2.1), i.e. we still have:

$$\frac{N_1}{N_2} = \frac{\omega_1}{\omega_2} = \frac{r_2}{r_1}$$

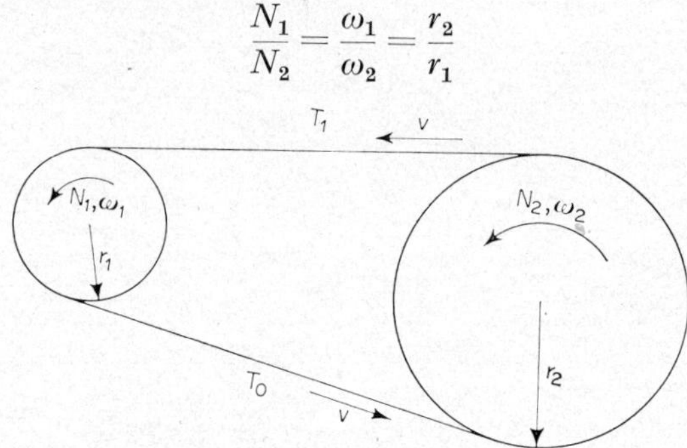

Fig. 2.5. Relative motion of pulleys in a belt drive.

Notice, however, that in this case the rotation of both rollers is in the same direction.

A belt drive is extremely versatile. By twisting the belt as shown in Fig. 2.6 it can be used to drive shafts that are not parallel, and in addition a single belt can be used to drive several shafts. Fig. 2.6 shows such a drive to the spindles of a spinning frame. The very narrow belt or tape passes round the four small

Fig. 2.6. Tape drive to spindles.

spindle pulleys in pairs before returning each time to the large driving roller. Because of the large ratio of diameters of the driving roller and the pulleys, the spindles are driven at high speed. The main purpose in driving more than one spindle from one tape is to maintain as even a tension as possible on the tapes. This ensures, as far as possible, that the spindles are all running at the same speed.

The fact that it is necessary to take special precautions to maintain an even tension suggests that the simple relationship given in Equation (2.1) does not apply exactly. The reason for this arises directly from the two assumptions that were made in order to apply this relationship to belt drives, namely that the belt is inextensible and that it does not slip on the pulleys. Neither of these assumptions applies exactly to belt drives. It is necessary for the belt tension to vary as it passes round the two pulleys if it is to transmit power, since the power transmitted is directly proportional to the difference in tension between the two sides of the belt.

The definition of work done is the product of the force times the distance through which the force is moved, and it follows therefore that the work done per unit time, i.e. the power, equals the product of the force times the speed of movement of the belt. If the tension, T_1, is in pounds force and the velocity, v, is in feet per minute, then the power transmitted from the pulley 1 to the pulley 2 is $T_1 v$ ft lbf/min (see Fig. 2.5). The power $T_0 v$ is transmitted back from the second pulley to the first so that the net power transmitted is:

$$(T_1 - T_0)v \text{ ft lbf/min}$$

Hence the horsepower transmitted is:

$$(T_1 - T_0)\frac{v}{33,000} \tag{2.2}$$

We have assumed that T_1 is greater than T_0, which will apply if pulley 1 is the driving pulley; if pulley 2 is the driving pulley T_0 will be greater than T_1, and power will be transmitted from the second pulley to the first.

The belt always increases in length under tension, and some of this extra length of the belt must be lost as the belt moves round from the high tension to the low-tensioned section. This results in a creeping of the belt round the pulley. It has been verified experimentally that this occurs, and that as a result the rule that the ratio of the rotational speeds of the two shafts should be inversely proportional to the pulley diameters is always found to be inaccurate, even in the most carefully designed belt drives. A further reason for the inaccuracy of speed transmission of belt drives is the possibility that the belt as a whole may slip on the pulley. In the next section the conditions under which such slipping can take place are discussed so that the design limitations of this type of drive can be appreciated.

2.4 CONDITIONS OF CRITICAL SLIPPING OF BELTS

It would appear at first sight that the difference between the input and output tensions of a belt which is on the point of slipping will be equal to μP, where

V-belt. The coefficient of friction between a flat belt and a pulley is very variable but a reasonable average value is 0·16. The actual coefficient of friction between a V-belt and the grooved pulley is about 0·2, and since the angle α (the groove angle) of the standard V-belt is 40°, the effective coefficient of friction is 0·2 cosec 20°, i.e. 0·614, a value more than three times as great as that for a flat belt drive.

To illustrate the size of the effect that these factors have on the power-carrying capacity of belts, let us consider a specific example. We shall compare the maximum horsepower transmitted per square inch by a flat belt and a V-belt to a pulley of 6 in radius, rotating at 150 r.p.m., the angle of wrap on each belt being 150°.

Consider the flat belt first. Since the maximum horsepower is required we put the output tension T_1 equal to the maximum load the belt will bear, namely 250 lbf/in². T_0 is then given by Equation (2.3):

$$\log_e \frac{250}{T_0} = \mu\theta = 0.16 \times \frac{150 \times \pi}{180} = 0.417$$

(Note that θ is expressed in radians, *not* degrees.)

Converting to logs to the base 10, using the equation $\log_{10} n = 0.4343 \log_e n$, we have:

$$\log_{10} \frac{250}{T_0} = 0.4343 \times 0.417 = 0.1811$$

Therefore:

$$\frac{250}{T_0} = 1.52$$

or

$$T_0 = 164 \text{ lbf}$$

The belt speed is

$$2\pi r N = 2\pi \frac{6}{12} 150 = 471 \text{ ft/min}$$

The maximum horsepower transmitted at 150 r.p.m. is therefore

$$(T_1 - T_0)\frac{v}{33,000} = (250 - 164)\frac{471}{33,000} = 1.23$$

If we now consider the V-belt, T_1 is 400 lbf. Hence:

$$\log_e \frac{400}{T_0} = \text{Effective value of } \mu \times \theta$$

$$= 0.2 \text{ cosec } 20° \times \frac{150\pi}{180}$$

$$= 1.62$$

so that

$$T_0 = 78.9 \text{ lbf}$$

The horsepower transmitted now becomes:

$$(400 - 78 \cdot 9)\frac{471}{33,000} = 4 \cdot 60$$

It can be seen that the V-belt can transmit nearly four times as much power as the flat belt.

The calculations given above are accurate only when the belt speed is relatively low, because at high speeds the belt tends to fly radially off the surface of the pulley, and a tension must be applied to both halves of the belt to keep it in contact with the pulley. Under these conditions Equation (2.3) is modified to:

$$\log_e\left(\frac{T_1 - T_2}{T_0 - T_2}\right) = \mu\theta$$

where T_2 is the tension required to keep the belt on the pulley.

Equation (2.3) has many other uses in the textile industry apart from predicting the ratio of belt tensions. It will also predict the ratio of tensions of a yarn or fibre as it slips past cylindrical objects such as tensioners, needles, etc. In using this equation, however, it should be remembered that it is based on the assumption that the rubbing materials obey Amonton's law. This states that the frictional force, F, is directly proportional to the load applied, P, i.e. that $F = \mu P$. For most textile fibres this law is not exactly obeyed, and it is more accurate to write:

$$F = \mu P^n$$

where n is a number which lies between $\frac{2}{3}$ and unity.

By assuming that n is very nearly unity—which is usually the case—it can be shown that for such a frictional behaviour the relation between T_1 and T_0 becomes:

$$\log_e\left(\frac{T_1}{T_0}\right) = \mu(\rho/T_1)^{1-n}\theta$$

where ρ is the radius of the pillar or pulley over which the yarn slides.

The basic advantages and disadvantages of belt drives arise from the frictional nature of the drive. Since the belt can slip it acts as a torque limiting device so that, when an electric motor is connected via a belt to a machine, the motor can be switched on without any fear that the sudden acceleration of the machine will result in breakage of some of its parts. The machine will always accelerate at a limiting value determined by the maximum torque that the belt will transmit. The frictional nature of the drive will, however, result in the speed ratio of the two pulleys of the drive being variable. Where small variations in speed are tolerable the belt drive is used widely, as, for example, on the main drive to most textile machines, but inside the machines it is usually necessary to ensure that the various motions are correctly synchronised, and for such purposes belts cannot be used. The frictional nature of belt drives does, however, make them very suitable for use in variable-speed drives, which we turn to next.

2.5 VARIABLE-SPEED FRICTION DRIVES

It is often necessary to provide variable-speed drives either to the whole, or to parts, of a textile machine, e.g. a circular sock or stocking machine must run at a lower speed when making the heel and toe than when making the rest of the sock or stocking. The simplest device for obtaining such a varying speed makes use of the fact that a belt can slide across from one set of pulleys to another set, and if the ratio of the diameters of these pulleys is different the speed of the driven shaft will be altered. This principle can be demonstrated most easily by considering the belt drive shown in Fig. 2.9. This consists of two cone-shaped pulleys connected by a belt. When the belt is in the position A the radius of the driven pulley is smaller than that of the driving pulley, while at the end B the opposite condition applies. Consequently, as the belt is moved across the two

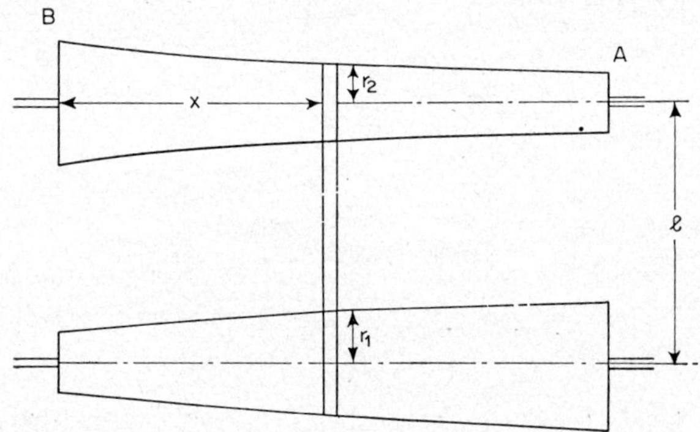

FIG. 2.9. Dimensions of a variable-speed belt drive.

pulleys the speed of the driven pulley changes continuously from a speed which is in most applications about twice that of the driving pulley to about half the speed of the driving pulley. The actual shape of the two cones must satisfy two conditions:

    1   That the length of the belt required at any position along the pulleys is constant.

    2   That the rate of change of speed ratio of the pulleys with position of the belt must satisfy some condition imposed by the purpose for which the variable speed mechanism is required.

To make this clearer let us consider a specific example. A cone flyer has two such cones, the sum of their radii at any belt position being equal to 4 in. It will be shown that such a condition will maintain constant belt length when the centre distance is large in comparison with the pulley radii—which applies to the cones in a cone flyer. To fulfil its control requirements the cones must be so shaped that, while the belt moves a distance of 2 ft parallel to the axes of the pulleys, the speed of the driven shaft varies linearly from 50 to 200 r.p.m., the speed of the driving pulley being 100 r.p.m. We are required to find the shape of the two pulleys. Since the two cones are relatively far apart it may be assumed that the angle of wrap around the two cones is the same, and equal to $180°$. It follows, therefore (see Fig. 2.9), that the length of the belt is given by:

$$2l + \pi(r_1 + r_2)$$

where l is the distance between the shaft centres, and r_1 and r_2 are the radii of the two pulleys at a given position of the belt. Since l is constant and the belt length must be constant, it follows that, at all points, $r_1 + r_2$ is a constant and equal to 4 in. In addition, we know that:

$$\frac{N_1}{N_2} = \frac{\omega_1}{\omega_2} = \frac{r_2}{r_1}$$

But from the control requirement, N_1/N_2 varies from $0\cdot5$ to $2\cdot0$ while the distance x of the belt from point B in Fig. 2.9 varies from 0 to 24 in. Hence:

$$\frac{r_2}{r_1} = \frac{N_1}{N_2} = 0\cdot5 + 1\cdot5\frac{x}{24} = 0\cdot5 + \frac{x}{16}$$

But

$$r_1 + r_2 = 4$$

hence

$$r_1 = \frac{64}{24 + x}$$

which is the required equation for the shape of the cone.

Using this equation the cone shapes shown in Fig. 2.9 have been obtained. If it were necessary to vary the speed according to some other function of the traverse, e.g. if the speed had to vary with the square of the distance moved by the belt, the equation connecting N_1/N_2 with x would be different but the principle of the determination of the cone shape remains the same.

This form of variable speed drive is often found in textile machinery. As has already been mentioned in the example given above, it is used to vary the bobbin speed in the cone flyer spinning system. The bobbin is fed with roving at a constant speed, and as the diameter of the bobbin builds up it is necessary to reduce the rotational speed of the bobbin to maintain a constant surface speed. Fig. 2.10 shows the cone pulleys used in a modern cotton roving frame. Fig. 2.11 shows a similar application on the Raper autoleveller draw frame. The purpose of this device is to vary the back roller speed of a gill box in synchronism with the weight per unit length of the input material, thus producing a much more even roving.

An even simpler application of the cone principle is shown in Fig. 2.12. This shows a positive yarn feed device for a knitting machine. This device must feed yarn into the machine at a constant rate, but this rate must be capable of being varied for different applications. The device shown in Fig. 2.12 consists of a simple cone covered with a high friction material, and a positioning device which controls the position where the yarn is fed onto and off the cone. The cone is driven at a constant speed and the delivery speed of the yarn is consequently determined by the position of the yarn on the cone.

FIG. 2.10. Cone pulleys on flyer drawing frame.

Another example of the same principle is shown in Fig. 2.13 which shows
the mechanism used for varying the speed of the warp beam of a warp knitting
machine. The rate of feed of yarn from the beam must be constant so that, as
the beam is used up and its diameter decreases, the rotational speed of the beam
much be increased. This is accomplished by the cone drive shown in Fig. 2.13,
in which view the second cone lies beneath that visible. Here, instead of connecting the cones by means of a belt, a ring is compressed between the two
cones. As the first cone rotates it forces the second cone to have the same surface
speed at the point where the ring is in contact with both cones. By moving the
ring the speed ratio is changed.

This form of direct frictional drive is also used in the device shown in
Fig. 2.14. The plate is driven by friction from the small wheel which is rotating
at constant speed. Since at the point of contact the plate and wheel must have
the same velocity (if they are not slipping), then

$$\frac{N_1}{N_2} = \frac{r_2}{r_1}$$

once more, where the angular speeds N_1 and N_2 refer to the wheel and plate
respectively, and r_1 and r_2 are the radii shown in Fig. 2.14. By moving the

wheel either towards the centre or towards the rim of the plate, it is possible to vary r_2 from a very small value to a value very much greater than r_1. By this means it is possible to obtain a very large speed variation. This device is rarely used except in calculating mechanisms since there is always slipping and wear because of the finite thickness of the wheel. This results in either the front or

FIG. 2.11. Cone pulleys on Raper autoleveller (by courtesy of Prince-Smith and Stells Ltd).

the back contact of the wheel slipping, since r_2 differs for these two sides of the wheel, and the speed condition supposed cannot apply to both sides.

An improved version of this drive which makes use of rolling rather than frictional contact is shown in Fig. 2.15. This figure shows the principle of operation of the Kopp variator. Two similar cones are attached to the input and output shafts, and these are connected by means of the rolling balls shown in

D

Fig. 2.15. When the axes of the balls are horizontal, as shown in Fig. 2.15a, the surface speed of the ball at the point of contact with the input shaft is the same as the surface speed at the same point of contact on the output shaft—as the distance of these two points from the axis of the ball is the same. The input and output speeds are therefore equal. When the balls are tilted as shown in Fig. 2.15b, the point of contact on the output shaft is much closer to the centre

FIG. 2.12. Positive feed cone for circular weft knitting machine (by courtesy of Hosiery and Allied Trades Research Association).

of rotation of the ball than the contact on the input shaft side; consequently the surface speed of the cone on the output shaft at the contact point is less than on the cone on the input shaft. The radius on the cone to the contact point is very little changed by this tilting of the ball, and consequently a reduction in rotational speed of up to 3 : 1 is produced by tilting the ball axes. Tilting the ball axes in the opposite direction, as shown in Fig. 2.15c, has the opposite effect and a speed increase of up to 3 : 1 can be obtained in this way.

A more robust form of variable speed drive is obtained by using a V-belt drive. The grooved pulley is replaced by two cone type flanges shown in Fig. 2.16. The driving pulleys consist of a flange fixed on the shaft and a flange

FIG. 2.13. Cone pulleys on warp beam speed control of
FNF warp knitting machine.

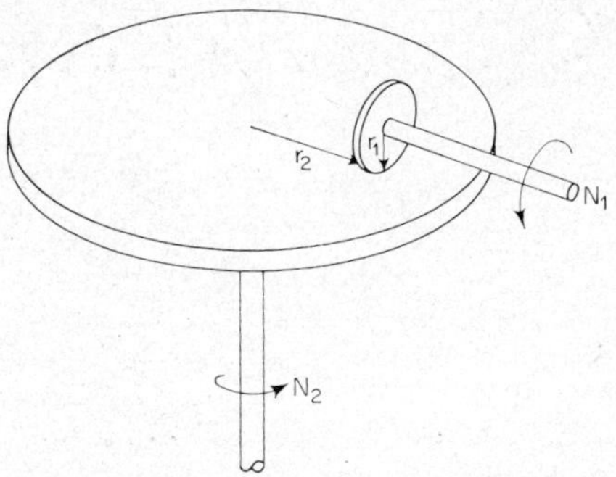

FIG. 2.14. Plate and wheel variable-speed drive.

which can be moved along the shaft axis. When the two flanges are close together the belt rides at the outside edge of the flange so that the effective radius

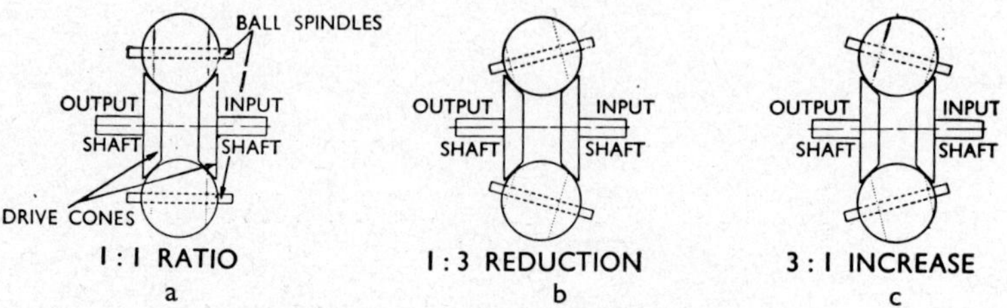

Fig. 2.15. Variable-speed drive of Kopp Variator
(by courtesy of Allspeeds Ltd).

of the pulley is large. As the adjustable flange is screwed away from the fixed
flange the V-belt moves down the side of the flange resulting in an effectively
lower pulley diameter. The flanges of the driven pulley also consist of one fixed
flange and a movable flange which is spring loaded. As the driving pulley is

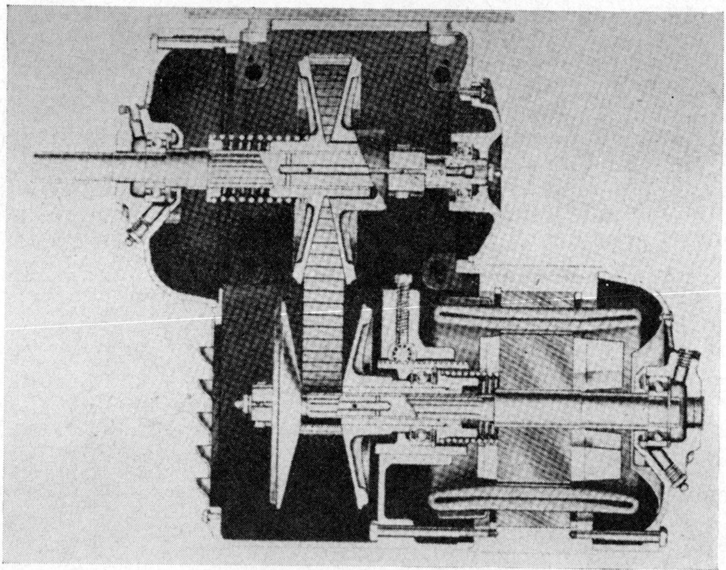

Fig. 2.16. Variable-speed V-belt drive
(by courtesy of U.S. Electrical Motors Inc.).

altered, say by moving the movable flange outwards, the driven flange moves
inwards, so increasing the effective diameter of the driven pulley. In this way
the speed ratio of the two pulleys is changed. The spring tension maintains the
belt tension at its proper value at all times.

To reduce slip even further, it is possible to replace the V-belt by a set of steel links held together by means of a chain. This is the principle of the PIV gear shown in Fig. 2.17. The links engage in grooves cut into the sides of the cone flanges so that the drive is positive and does not rely on friction.

FIG. 2.17. The PIV variable-speed drive
(by courtesy of The Link Belt Company).

Suggested Further Reading

T. BEVAN. *The Theory of Machines*. chapter 8. Longmans, 1943.
A. H. NISSAN (Editor). *Textile Engineering Processes*, chapter on mangles. Butterworths, 1959.

CHAPTER 3

THE TRANSMISSION AND MODIFICATION OF ROTATIONAL MOTION—POSITIVE DRIVES

3.0 INTRODUCTION

THE final example of variable-speed drives mentioned in the last chapter has introduced the concept of a more positive drive. It has already been noted at the end of Section 2.3 that more positive drives are necessary if correct synchronism is to be maintained between different parts of a machine. Because positive drives do not require such relatively large surfaces to transmit the driving force by friction they also produce more compact systems, especially when large powers are to be transmitted. There are three basic types of positive drive:

1 Roller chain drives.
2 Timing belts.
3 Gear drives.

The last-named is probably the most important.

3.1 ROLLER CHAIN DRIVES

A chain drive such as that shown in Fig. 3.1 consists essentially of a set of links, joined together by pin joints to form a chain, which are engaged by means of two sprockets. The distance between the two shafts is not critical for this type

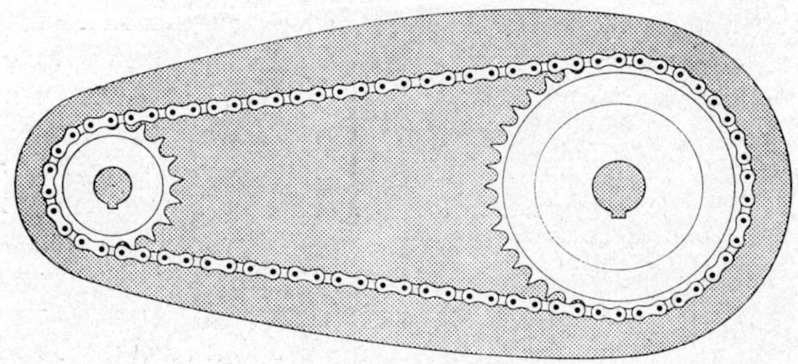

FIG. 3.1. Chain drive.

of drive, and it is particularly useful for driving shafts which need to be adjusted in position, or where the drive has to be conveyed to many shafts. Examples of such applications are shown in Fig. 3.2 and Fig. 3.3. Fig. 3.2 shows a chain drive to the intermediate rollers of a drafting frame. These rollers need to be

FIG. 3.2. Chain drive to intermediate rollers on drafting frame
(by courtesy of Prince-Smith and Stells Ltd).

FIG. 3.3. Chain drive to carding workers and strippers.

moved nearer to or further from the front rollers if fibres of different average lengths are to be processed. Fig. 3.3 shows the chain drive to the workers and strippers of a card. In this application the drive must be conveyed from the main shaft to as many as 12 other shafts which are some distance away. All the shafts must run in synchronism to maintain the correct carding action at all times. The chain drive forms the most suitable means for conveying the drive to each of these shafts.

The chain drive does possess a disadvantage that could be serious in certain applications. If the sprocket is driven at constant speed, then the chain will have a constant average linear speed. There will, however, be a small variation in the chain speed over the short interval from the time when the chain link is being engaged to its disengagement by any tooth of the sprocket. Figs. 3.4a and b show the positions of the chain and sprocket at two different stages of engagement. If the sprocket is rotating at N r.p.m. then point A on the chain will have a surface speed of $v = 2\pi r'N$ in the position shown in Fig. 3.4a, while $v = 2\pi rN$ in Fig. 3.4b. The difference between the radii r and r' will vary with the number of teeth on the sprockets. For normal sprockets this difference will

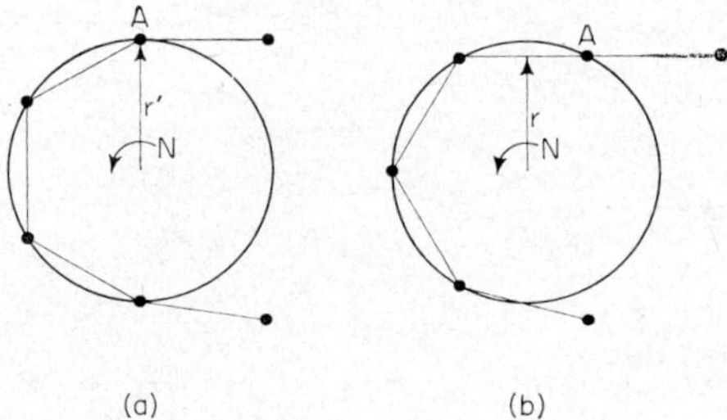

FIG. 3.4. Chain drive in two positions of engagement.

result in a variation in surface speed of the chain of 1 to 2 per cent. In the applications quoted above such a small variation in speed is unimportant, but if we were to attempt to drive the front and back rollers of a drafting system by means of a roller chain drive such variations in speed cannot be tolerated, as they would produce quite unacceptable variations in sliver or yarn thickness. In addition, this variation in speed will result in changes in tensions in the chain because of the need to accelerate and decelerate the mass of the driven part of the mechanism. This variation in tension will become more severe as the speed is increased. Chain drives are therefore limited to relatively low speed drives. For medium speeds, specially shaped links have been developed for a type of chain drive known as a silent chain. These shaped links tend to reduce the variation in chain speed, but do not eliminate it completely.

3.2 TIMING BELTS

A relatively new type of drive, the timing belt, is capable of high-speed positive drive. Fig. 3.5 shows a typical timing-belt drive. The link between the pulleys consists of a neoprene belt with neoprene teeth moulded on to the inside of the belt. The core of the belt is reinforced with steel wire or fibreglass to limit the stretch of the belt to a very small amount. The neoprene teeth mesh into teeth cut into the pulleys, thus providing a positive drive. These belts can be used

FIG. 3.5. Timing belt (by courtesy of Crofts (Engineers) Ltd).

for speeds up to 16,000 ft/min, are silent, long lasting, relatively cheap, and provide a very closely controlled means of transmitting rotational speed. They are being used for an increasing number of applications, in which they have replaced gears, but relatively little is known of their behaviour because of their novelty.

3.3 GEAR DRIVES

Of all the positive drives available for driving shafts which are relatively close together, gear drives are the most versatile and accurate form of drive. They can transmit a drive between shafts which are parallel or at an angle to each other, and they can operate at low or very high rotational speeds. Their efficiency is very high and the accuracy of transmission of speed is only dependent on the accuracy with which the gears can be made.

There are many different types of gearing and Fig. 3.6 shows the most important types. These six types represent two examples of each of three major groups of gears, each group having a particular relationship between the driving and driven shaft as shown in the accompanying table.

Table 3.1

Group	Shafts	Gear Example	
I	Parallel	Spur	(Fig. 3.6a)
		Helical	(Fig. 3.6b)
II	Intersecting at some angle to each other	Bevel	(Fig. 3.6c)
		Spiral bevel	(Fig. 3.6d)
III	Non-intersecting but at some angle to each other.	Worm	(Fig. 3.6e)
		Skew	(Fig. 3.6f)

From Fig. 3.6 it can be seen that each type of gear has in addition its own particular shape of tooth. Spur and bevel gears are straight gears cut on to a cylinder and frustum of a cone respectively, the teeth lying in the direction of the shaft axis. Helical and spiral bevel gears are curved or at an angle to the shaft axis. Worm gears are, on the other hand, merely a particular example of skew gears where the shafts are at right-angles to each other. Skew gears can in general have shafts at any angle to each other.

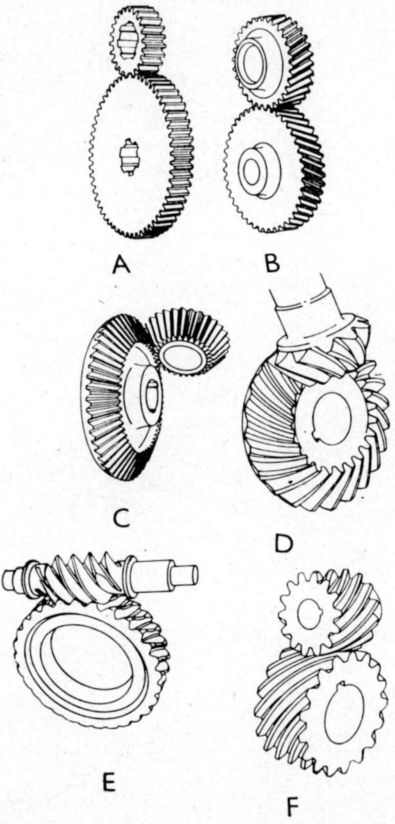

FIG. 3.6. Various forms of gear drives
(by courtesy of David Brown Corporation (Sales) Ltd.)

In all gear drives, the drive is transmitted from one shaft to the other by the teeth in the two gears meshing with each other. The form of meshing found in these gears varies considerably from one gear type to another. As will be shown later, the contact between spur and bevel gear teeth is basically a rolling contact while the contact between worm and skew gear teeth is a sliding contact. While a rolling contact results in less wear and reduced friction at the gear faces, it has the disadvantage that when the teeth engage and disengage, the rolling contact can result in vibration and noise in the gear transmission, which limits the use of spur and bevel gears at high speeds. The helical and spiral bevel gears have a geometry which results in a mixture of sliding and rolling contact, and are an attempt at taking advantage of both forms of contact; they can be regarded as the high-speed versions of the spur and bevel gears. They are comparatively rare in textile machinery, and as in addition they have a relatively complex geometry, they will not be considered any further here. The reader is referred to the bibliography on gearing for further information about helical gears.

3.4 SPUR GEARING

Spur gears are undoubtedly the most common form of gearing, and we shall consider their properties in some detail. A very important requirement of gears is that when two gears mesh they will behave as far as possible as if they were two cylinders rolling on each other, so that if the driving gear is rotating at a constant speed the driven gear will also rotate at a constant speed. To obtain this desirable condition the teeth on the gear wheels must be shaped in such a way that they roll smoothly on each other. For the moment we shall suppose that this has been done, and return to the question of the geometry of the teeth in a later section.

Fig. 3.7 shows two gears intermeshing, together with the principal dimensions of the gear, and the common names given to these dimensions. It can be seen that, since each tooth on one gear meshes with a tooth on the other, the rotational speeds of the two gears, N_1 and N_2, must be connected by the fact that the number of teeth that are brought into mesh per minute is the same for both gears, i.e.

$$N_1 n_1 = N_2 n_2$$

or
$$\frac{N_1}{N_2} = \frac{n_2}{n_1}$$

where n_1 and n_2 are the number of teeth on each gear;

and clearly, also
$$\frac{\omega_1}{\omega_2} = \frac{n_2}{n_1} \tag{3.1}$$

If we consider the two gears to be equivalent to two circles of diameters d_1 and d_2 rolling on each other, then we have

$$\frac{d_2}{d_1} = \frac{N_1}{N_2} = \frac{n_2}{n_1}$$

But since the distance between the gear centres is known the sum of the two diameters is fixed. Hence:

$$d_1 + d_2 = D, \text{ say,}$$

and therefore

$$d_1 = \frac{D}{\left(1 + \dfrac{n_2}{n_1}\right)} \text{ and } d_2 = \frac{D}{\left(1 + \dfrac{n_1}{n_2}\right)} \tag{3.2}$$

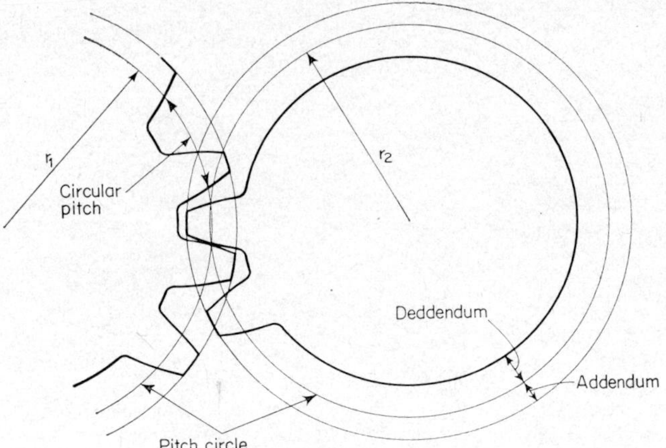

FIG. 3.7. Principal dimensions of two meshing gears.

Equations 3.2 specify the diameters of the two circles, and these are shown on Fig. 3.7 as the pitch circles of the two gears. From the above argument, also:

$$\frac{n_2}{d_2} = \frac{n_1}{d_1}$$

This implies that if two gears are to mesh, the ratio of the number of teeth to the pitch diameter, i.e. (n/d), must be the same for both gears. This ratio is rather inappropriately known as the diametral pitch (DP) and has units of inches^{-1}. It will be shown in the next section that the pitch circle must lie somewhere between the top and bottom of the tooth, and since teeth of varying addenda or deddenda (i.e. lengths of tooth above or below the pitch circle— see Fig. 3.7) can still mesh correctly, it is evident that the vital distance which determines the ability of teeth to mesh correctly is the distance along the pitch circle between neighbouring teeth on the same gear. This distance is known as the circular pitch and is clearly related to DP, as defined above. From the definition of circular pitch we see that:

$$\text{Circular pitch} = \frac{\pi d}{n}$$

$$= \frac{\pi}{\text{DP}}$$

We have seen above that if two gears are to mesh correctly the DP of the two gears must be the same. The DP is therefore used to standardise gears. The circular pitch could be used for this purpose but since the number π is involved in its definition it is not very convenient for this purpose. An advantage of the diametral pitch is that its value is always a whole number or a fraction. To make this clearer let us consider a specific example.

If a 30-tooth gear meshes with a 60-tooth gear, both gears being 24 DP, let us calculate the distance between the gear centres, and the circular pitch of the two gears. By definition:

$$DP = \frac{n}{d}$$

or

$$d = \frac{n}{DP}$$

For the first gear with 30 teeth, we therefore have:

$$d_1 = \frac{30}{24}$$

while for the second gear, with 60 teeth

$$d_2 = \frac{60}{24}$$

Since the pitch circles roll on each other, the distance between the centres is half the sum of the two diameters (notice the similarity to the example on p. 20). Hence the centre distance is $\frac{1}{2}(60 + 30)/24 = 1\frac{7}{8}$ in. The circular pitch for both gears is $\pi/24$ in.

It should be noted that coarse-toothed gears have a diametral pitch which is small, but have a large circular pitch, and vice versa for fine-toothed gears.

A further quantity which is of importance in considering the shape of gear teeth is the module. This is defined as the circular pitch divided by π; i.e.

$$module = \frac{circular\ pitch}{\pi}$$

$$= \frac{d}{n}$$

$$= \frac{1}{DP}$$

It is usual to make the addendum of a gear equal to the module (some of the reasoning behind this rule will be discussed later). If we assume that this rule is followed, what is the outside diameter of the two gear wheels previously considered? Clearly, the outside diameter is equal to the diameter of the pitch circle plus two modules. Hence the outside diameters are:

$$\frac{30}{24} + \frac{2}{24} = 1\frac{1}{3}''$$

and

$$\frac{60}{24} + \frac{2}{24} = 2\frac{7}{12}''$$

It should be noted that half the sum of the outside diameters of the gears is clearly greater than the centre distance of the gears.

3.5 THE SHAPE OF GEAR TEETH

In the previous section the general dimensions of gear teeth have been discussed, but their actual shape has not been considered. The shape of gear teeth are chosen so as to ensure that if the driving gear is rotating at a constant speed the driven gear is also rotating at a constant speed. It is obvious from the discussion of the previous section that if two gears have the same values for DP and mesh correctly, the ratio of their *average* rotational speeds is constant, since this ratio is equal to the ratio of the numbers of gear teeth on the two gear wheels. But as we have already seen, when considering the speed variations of a roller chain drive, it is possible that the instantaneous speed will vary above and below this average speed as the gear teeth move into mesh and out again. This seems at first sight to be inevitable because when the gears begin to mesh the contact point on the driven gear will lie at the tip of the tooth where the radial distance is larger, and will gradually move down the tooth as the teeth mesh more deeply (see Fig. 3.7). Consequently the radial distance to the contact point will get smaller as the teeth mesh, and the reverse will happen on the driving gear.

This variation in the radial distance of the contact points during meshing would imply a variation of rotational speed of the driven gear if the linear speeds of the two contact points between two gear teeth were the same. The two contact points do not have the same linear speed but they *do* have the same *component* of speed in the direction of the common normal to the contacting surfaces, since they are in contact. (This direction is at right-angles to the common tangent to the contacting surfaces.) The total linear speed of each contacting point is, however, always at right-angles to the radius to that point (since the gear as a whole is rotating). This direction is not, in general, the same for both gears, nor does it necessarily lie in the direction of the common normal to the surfaces. As a result it is not necessary for the linear speed of the two contacting points to be the same. It is therefore possible to design the shapes of gear teeth in such a way that the *rotational* speed of the points of contact is not only the same on average, but is always instantaneously the same. To do this, use is made of a simple geometrical theorem which we now describe.

Fig. 3.8 shows two gear teeth in contact at the point A. O_1 and O_2 are the centres of the two gear wheels, and we shall suppose that the angular velocities of the two gear wheels are ω_1 and ω_2 respectively. The contact point A can be thought of in two ways.

Firstly, it can be regarded as a point on gear 1. In this case it is a point rotating about the centre O_1 with an angular speed ω_1. Its actual velocity is therefore at right-angles to the line O_1A and has a magnitude $\omega_1.O_1A$. In the same way, A can be thought of as a point on gear 2, rotating about O_2 with angular velocity, ω_2. In this case A has an actual velocity at right-angles to O_2A, of magnitude $\omega_2.O_2A$. These velocities are shown on Fig. 3.8.

Suppose the line C_1PAC_2 is the common normal to the two teeth surfaces at the contact point A, and draw O_1C_1 and O_2C_2 perpendicular to this common normal. Now O_1C_1 is perpendicular to AC_1, and O_1A is perpendicular to the velocity of A regarded as a point on gear 1. Hence the velocity $\omega_1.O_1A$ of A makes the same angle θ_1 with the common normal as does O_1A with O_1C_1. The

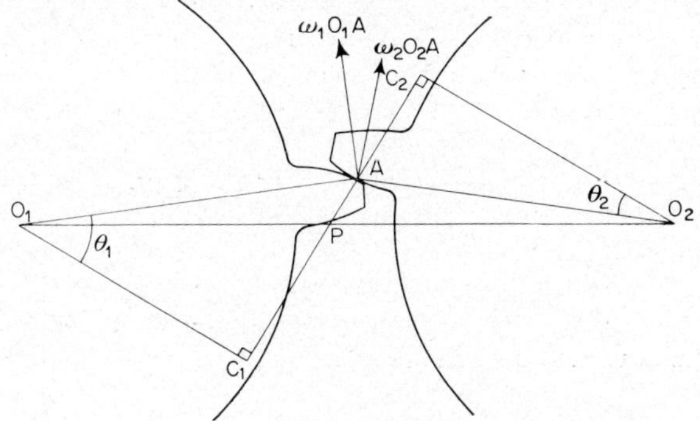

FIG. 3.8. Velocities of contact point of two gears.

component of the velocity of A on gear 1 along the common normal is therefore:

$$\omega_1.O_1A \cos \theta_1$$

In the same way it can be shown that the velocity component of A on gear 2 along the common normal is:

$$\omega_2.O_2A \cos \theta_2$$

But we have seen that if the teeth are to remain in contact, these two velocity components must be the same, i.e. we must have

$$\omega_1.O_1A \cos \theta_1 = \omega_2.O_2A \cos \theta_2$$

or

$$\frac{\omega_1}{\omega_2} = \frac{O_2A \cos \theta_2}{O_1A \cos \theta_1}$$

Now, since O_1C_1A is a right-angle, we see from triangle O_1C_1A that:

$$O_1C_1 = O_1A \cos \theta_1$$

Similarly

$$O_2C_2 = O_2A \cos \theta_2$$

and therefore

$$\frac{\omega_1}{\omega_2} = \frac{O_2C_2}{O_1C_1}$$

Let P be the point where the line of centres O_1O_2 meets the common normal C_1PAC_2. Then the triangles O_1PC_1 and O_2PC_2 are similar, so that

$$\frac{O_2C_2}{O_1C_1} = \frac{O_2P}{O_1P}$$

Consequently:

$$\frac{\omega_1}{\omega_2} = \frac{O_2P}{O_1P} \tag{3.3}$$

Now, if ω_1/ω_2 is to be constant, the ratio O_2P/O_1P must be a constant. But $O_1P + O_2P$ is equal to the fixed distance between the centres of the gear wheels, i.e.

$$O_1P + O_2P = \tfrac{1}{2}(d_1 + d_2) \tag{3.4}$$

where d_1 and d_2 are the diameters of the pitch circles of the two gears. If n_1 and n_2 are the number of teeth on the two gears, and if the DP is the same for each, then from the definition of DP:

$$\frac{n_1}{d_1} = \frac{n_2}{d_2}$$

or

$$\frac{n_1}{n_2} = \frac{d_1}{d_2}$$

Now from Equation (3.1) we know that

$$\frac{\omega_1}{\omega_2} = \frac{n_2}{n_1}$$

$$= \frac{d_2}{d_1}$$

Therefore

$$\frac{O_2P}{O_1P} = \frac{d_2}{d_1} \tag{3.5}$$

Solving Equations (3.4) and (3.5) for O_1P and O_2P we find that:

$$O_1P = \tfrac{1}{2}d_1 \quad \text{and} \quad O_2P = \tfrac{1}{2}d_2$$

Therefore P is a fixed point lying on the pitch circles of both gears. This point is known as the pitch point.

Thus, what this theorem states is that if two gears are to mesh correctly, i.e. in such a way that both gears always rotate at a constant speed, the teeth must be shaped so that their common normal at the point of contact always passes through the fixed pitch point.

This is a very important conclusion because it means that if the teeth on one of the gears have any arbitrarily chosen shape there is, in general, only one tooth shape that can be used on the other gear if the gears are to mesh correctly. Two gear shapes that mesh correctly with each other are said to be conjugate.

As an extreme example of what can be achieved the gear teeth in some watch and clock mechanisms are so designed that they will mesh correctly even though one of the gears consists of a set of pins set into the circumference of a wheel. However, such conjugate pairs are designed for very special purposes, and it is obviously an advantage to have sets of gears any one of which will mesh correctly with any other. This condition would be satisfied if the tooth shape was such that its conjugate was the same shape. This can be done, and the most important of the shapes that are conjugate with themselves is called the involute, which will be considered in the next section.

Before doing so, however, it is important to note that there is an important corollary to the theorem we have just proved. If the gear shapes are conjugate and produce constant rotations of the gear wheels when new, it follows that this will no longer be the case when the gears have worn and their shape has altered. Worn gearing will result in a speed variation during each meshing and unmeshing cycle of the gears, so that a relatively rapid fluctuation in speed of the driven gear will arise. This will occur n times during a single rotation of the gear, if n is the number of teeth on the gear. The frequency of this speed variation is of considerable importance in many textile applications where gears are used to drive take-up rollers, as, for example, in drafting frames and looms. Variations in speed of the take-up rollers result in irregularities in the yarn or cloth being produced. If these irregularities occur at very frequent intervals they can often be ignored, and the gears used are consequently chosen to ensure that the frequency of the gear speed variation is sufficiently high. To show how this is achieved we shall return to this topic when the behaviour of gear trains has been considered.

3.6 THE INVOLUTE GEAR

Imagine a piece of string wound round the circumference of a cylinder. If the string is unwrapped from the cylinder, always keeping the string taut, then the curve traced out by the end of the string is called an involute (see Fig. 3.9). This shape has an interesting property, namely, that the normal to the curve always lies along the string. This follows directly from the fact that the velocity

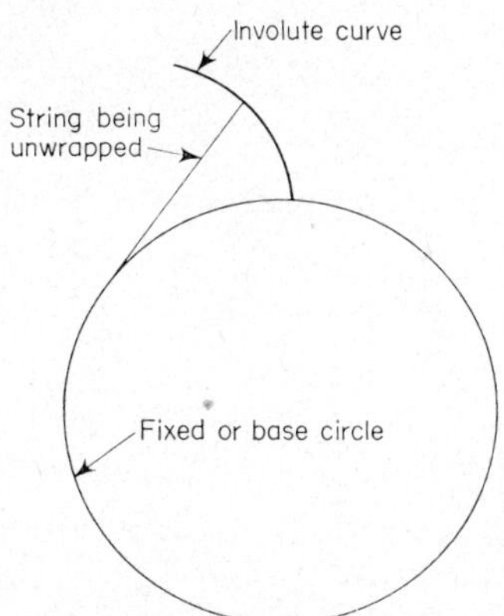

Fig. 3.9. Properties of involute.

of the end of the string, which must lie along the tangent to the curve, is always at right-angles to the string.

Consider now two cylinders with centres O_1 and O_2, and radii r_1 and r_2. Round these cylinders is wrapped an endless string which crosses over itself as it passes from one cylinder to the other, as shown in Fig. 3.10. Suppose a pencil is attached to the string at a point A on it. Now imagine the cylinder 1 rotating in a clockwise direction, and suppose that, initially, the pencil A was at the point where the string leaves cylinder 1 at point A_1 (see Fig. 3.11). Then A

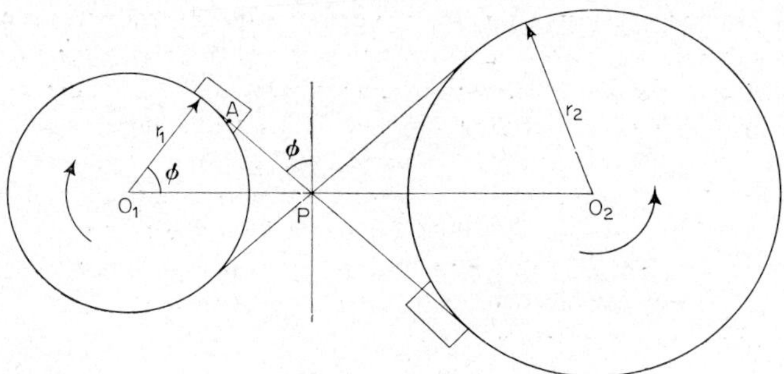

FIG. 3.10. Production of involute teeth—general view.

can be thought of as the end of a taut string being unwound from the cylinder. In other words, if a board is fixed to the cylinder in the vicinity of A, the pencil will trace an involute on this board, as shown in Fig. 3.11. Eventually the pencil

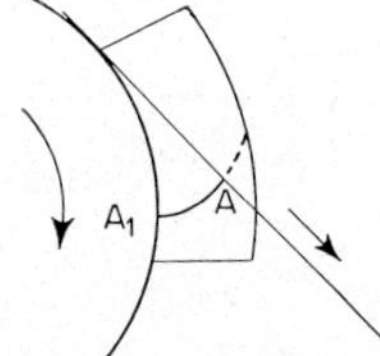

FIG. 3.11. Production of first involute tooth.

A will leave the board on cylinder 1, and move towards cylinder 2. We can now think of A as being the end of a taut string that is being wound onto the second cylinder. It will therefore trace an involute on a board fixed to this cylinder, as shown in Fig. 3.12.

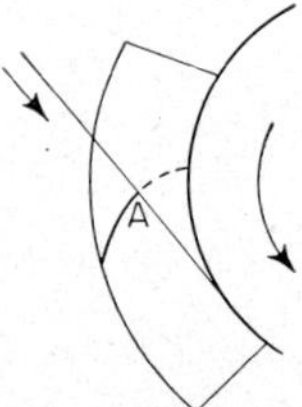

FIG. 3.12. Production of mating involute tooth.

It is obvious that the normals to these involutes must lie along the line of the string and therefore must pass through the point P where the strings cut the line of centres. Therefore if gear teeth were made with their sides in the shape of these involutes the teeth would mesh correctly, since the normals to them at the point of contact would always pass through the fixed point P, which would be the pitch point.

Furthermore, in the absence of friction the forces exerted by one gear tooth on the other at the point of contact must also always act along the line of the string. It can be shown that the involute gear shape is the only one that has this property of constant direction of inter-gear force, and as will be shown later this fact is a considerable advantage in practice. In fact, it has resulted in involute gears becoming practically the standard gear shape for all spur gears.

Before considering the advantages of this particular shape it is as well to note certain facts with regard to the involute shape. To construct an involute gear it is necessary to know the circle sizes from which the involute is formed. The two circles shown in Fig. 3.10 are known as the base circles of the involute. They clearly have a smaller radius than the pitch circles, which pass through the pitch point P. The ratio of the base circle r_1 to the pitch circle is given by r_1/OP, and from Fig. 3.10 it can be seen that:

$$r_1 = \text{(pitch circle radius)} \times \cos \phi = \tfrac{1}{2}d_1 \cos \phi$$

where ϕ is the angle shown in Fig. 3.10.

The angle ϕ is known as the *pressure angle*; it is the angle between the line of action of the force between the gear teeth at their point of contact and the perpendicular to the line of centres (see Fig. 3.13). Since this line of action lies along the normal to the gear profiles it will, for all correctly designed gears, pass through the pitch point P, but the involute gear is the only gear for which the pressure angle is constant throughout the whole time the gears are engaged.

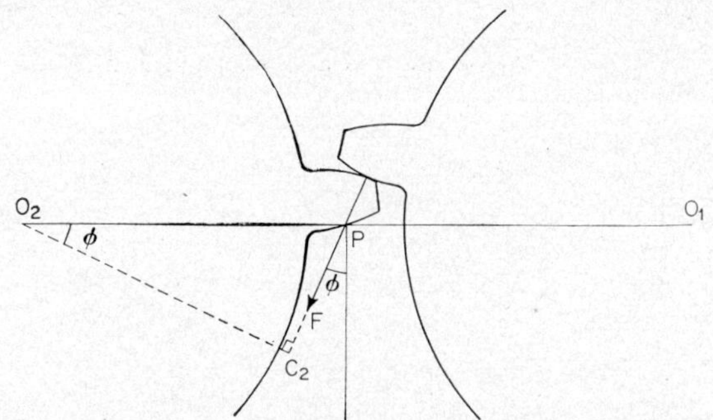

FIG. 3.13. Line of pressure between two involute teeth.

Referring to Fig. 3.13 we see that the average torque on gear 2 is

$$F \times O_2C_2 = \tfrac{1}{2}Fd_2 \cos \phi$$

where F is the force between the two gear teeth. Similarly, the average torque on gear 1 is

$$\tfrac{1}{2}Fd_1 \cos \phi$$

Also, the load on the bearings at O_1 and O_2 is equal to the component of F along the line of centres, i.e.

$$\text{Load on bearings} = F \sin \phi$$

Clearly, it is desirable to have as small a load on the bearings as possible, and therefore ϕ should be kept small when designing the gear. However, if ϕ is made too small the general effectiveness of the gear is limited. This point will be discussed more fully in the next section. In practice, a working compromise is given by the British Standard specification 436, which suggests a standard value of $\phi = 20°$.

Since the involute gear is the only one that has a constant pressure angle, it is the only one that can maintain a constant loading on the bearings. This is clearly of advantage in limiting the vibration of the gears during running.

A further advantage arises from the shape of the gear tooth when the gear wheel is very large. If the gear circle becomes very large the shape of the tooth will approximate to a straight line. In fact, for a gear cut on to a straight bar (which can be considered to be a section of an infinitely large gear wheel) the profile of the gear consists of two straight lines at an angle ϕ to the vertical to the bar. A gear section like this, which is known as a gear rack, can clearly be made very accurately. If such a rack is made with cutting edges to the 'gear' teeth, it can be made to cut out involute teeth by 'meshing' the gear blank with the rack, the rack moving forward and the gear blank being rotated at the appropriate speed as if the gears were meshing while cutting takes place. The cutting edges will then generate the gear shapes. This is a relatively simple method of producing accurate involute gear teeth.

The involute gear also has the unique property that if two gear wheels are meshing correctly at their standard distance apart (i.e. at a distance equal to the sum of the radii of their pitch circles), they will still mesh correctly when the centre distance of the gears is increased. This follows directly from the fact that the same involute is produced in Fig. 3.10 if the two circles are moved further apart, so long as the base circles are kept the same. Since the rotational motion of the gears shown in Fig. 3.10 can be considered to be the same as that of the two base circles connected by an endless belt it is evident that the speed ratio is constant and independent of the distance apart of the centres. If the gears are moved further apart the pitch circle diameters, and consequently the DP, together with the pressure angle ϕ, are all altered. Because the pressure angle, DP, etc., change when the distance between the centres is changed, their standard values are defined when the centre distance is such as to produce the standard pressure angle, $\phi = 20°$. If this distance is altered the teeth will not mesh as deeply but they will still mesh correctly even though the only geometrical relationship which has remained the same is the size of the two base circles.

The advantages of the involute have made such gears almost universal. The involute shape has, however, one disadvantage which we shall consider in the next section. Before doing so we give now an example of the type of calculations involved in designing gears.

Two involute gears, of 24 DP and $20°$ pressure angle, are in mesh. One gear has 30 teeth, the other has 120 teeth, and the torque on the smaller gear is 20 lbf in. We shall calculate the size of the base circles, the distance between the centres of the gears, and the pressure on the gear bearings.

DP is defined as n/d, where n is the number of teeth on the gear and d is the diameter of the pitch circle. Thus, we have:

$$d_1 = \frac{30}{24} = 1 \cdot 25$$

and

$$d_2 = \frac{120}{24} = 5 \cdot 00$$

The distance between the gear centres is therefore:

$$\tfrac{1}{2}(d_1 + d_2) = \tfrac{1}{2}(1 \cdot 25 + 5 \cdot 00) = 3 \cdot 125 \text{ in}$$

To find the radii of the base circles, we use the relation

$$r = \tfrac{1}{2}d \cos \phi$$

Thus

$$r_1 = \tfrac{1}{2} \times 1 \cdot 25 \times \cos 20° = 0 \cdot 587 \text{ in}$$

and

$$r_2 = \tfrac{1}{2} \times 5 \cdot 00 \times \cos 20° = 2 \cdot 350 \text{ in}$$

The torque on the smaller gear (gear 1) is known to be 20 lbf in. Hence:

$$\tfrac{1}{2}Fd_1 \cos 20° = 20$$

which gives

$$F = 34 \text{ lbf}$$

Therefore the load on the bearings is

$$F \sin \phi = 34 \sin 20° = 11 \cdot 6 \text{ lbf}$$

A quantity of great interest is the shortest distance between the base circles. From Fig. 3.10 this can be seen to be equal to:

$$\tfrac{1}{2}(d_1 + d_2) - (r_1 + r_2)$$
$$= \text{distance between centres} \ - (r_1 + r_2)$$
$$= 3 \cdot 125 - 2 \cdot 937$$
$$= 0 \cdot 188 \text{ in}$$

The calculation shows that this distance is relatively small. Clearly the involute tooth height cannot be greater than this distance, which becomes smaller if ϕ is decreased. This is the main reason why ϕ cannot be made too small, and we must now examine more closely the factors that determine the total height of the tooth.

3.7 THE GEAR TOOTH HEIGHT

Two necessary conditions for the correct meshing of two gear wheels have already been mentioned, namely:

    1 If the rotational speed of the driving gear is constant, the rotational speed of the driven gear should also be constant.

    2 The gears must have the same DP.

In addition, two further conditions must be fulfilled if the gears are to operate correctly. These are as follows:

3 In order to provide a continuous drive from one gear to the other, at least one tooth on the driving gear must be in contact with a tooth on the driven gear.

4 As the teeth mesh the contact point changes with time. Therefore to ensure that condition 3 holds at all times, there must be instants when more than one pair of teeth are in contact. It is essential that the contact of the second pair does not affect in any way the contact of the first pair, i.e. there must be no interference between the pairs of teeth in contact.

Whether or not condition 3 is fulfilled depends primarily on the height of the gear teeth. Fig. 3.14 shows the base circles of two involute gears and the line of contact of two meshing gear teeth, i.e. the common tangent to the base circles. Also shown are two different addendum circles on each gear. These circles are drawn through the topmost points of the gear teeth. The addendum circles on gear 1 cut the line of contact at points E_1 and E_2, and these circles on gear 2 cut it at F_1 and F_2. The points E, F on any one size of addendum

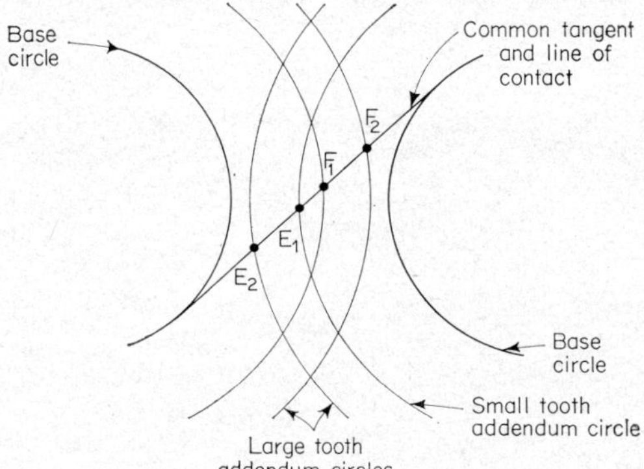

FIG. 3.14. The effect of tooth size on meshing length.

circle clearly represent the beginning and end of gear tooth contact, and it can be seen from the diagram that the contact length (i.e. the distance EF along the contact line) is increased as the addendum is increased. To ensure that there is sufficient contact length between the gears, it is common practice to make the addendum equal to the module (i.e. the reciprocal of DP). Under these conditions the average number of pairs of teeth in contact is about two.

If the gear tooth were to extend only as far as the base circle (the involute shape has no meaning below the base circle), then the above requirement will necessarily limit the lowest number of teeth that can be used on a gear. The addendum of the mating tooth on a large gear wheel must be less than the distance between the pitch circle and the base circle on the second (smaller) gear, since otherwise the top of the tooth will meet the solid blank of the small gear (see Fig. 3.7).

As we have seen, the base circle diameter is $d \cos \phi$, where d is the diameter of the pitch circle. Hence the radial distance between the pitch and base circle is:

$$\tfrac{1}{2}(d - d \cos \phi) = \tfrac{1}{2}d(1 - \cos \phi)$$

If the addendum is equal to the module, m, we then have

$$\tfrac{1}{2}d(1 - \cos \phi) > m$$

But $m = 1/DP$, and $d\,DP$ equals the number of teeth on the gear; hence:

$$\tfrac{1}{2}(1 - \cos \phi) > \frac{1}{n}$$

or

$$n > \frac{2}{1 - \cos \phi}$$

For a pressure angle of 20° this condition requires n to be greater than 34. This restriction on the minimum number of teeth is found to be too restrictive in practice, and standard involute gear teeth do in fact extend below the base circle. No contact should take place within the region below the base circle because no correctly meshing section can be cut there. This section is merely cut away to allow the teeth to penetrate below the base circle of the opposing gear. The standard shape consists of two radial lines below the base circle extending far enough to produce a dedendum of 1·157 modules. The shape chosen does not result in any contact between the involute and the portion below the base circle if the number of gear teeth is again above a certain minimum.

The proof of this fact requires a straightforward but lengthy geometrical argument for which the reader is referred to any standard text on gear dimensions. Suffice it to say here, that for a pressure angle of 20° a minimum of 17 teeth is necessary, and for a pressure angle of $14\tfrac{1}{2}°$ the minimum is 32 teeth. Since the larger pressure angle produces a larger possible range in the number of teeth, and also results in stronger teeth, it has been preferred, despite the larger bearing loads that result from its use.

This limitation on the minimum number of teeth on any gear wheel applies

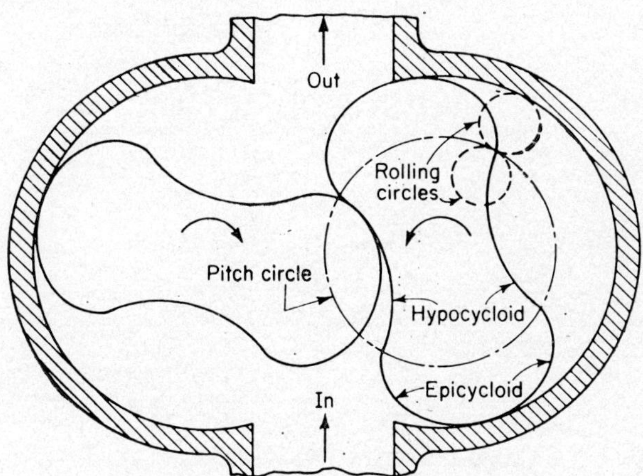

Fig. 3.15. Cycloidal gear—Roots blower.

specifically to involute gears. A gear shape which does not possess this disadvantage is the cycloidal gear shape. Fig. 3.15 shows two cycloidal gears meshing, and the number of teeth on these gears is only two. Such gears are, however, used only for very specific purposes, and, in general, one can say that gear wheels cannot have less than about 17 teeth.

3.8 THE EFFICIENCY OF SPUR GEARS

When two gears are meshing and the point of contact lies along the line of centres the relative motion of the two gear teeth is one of pure rolling. This can be seen by looking at Fig. 3.16 and noting that at this point the velocity of both contacting surfaces is at right-angles to the line of centres. The points of

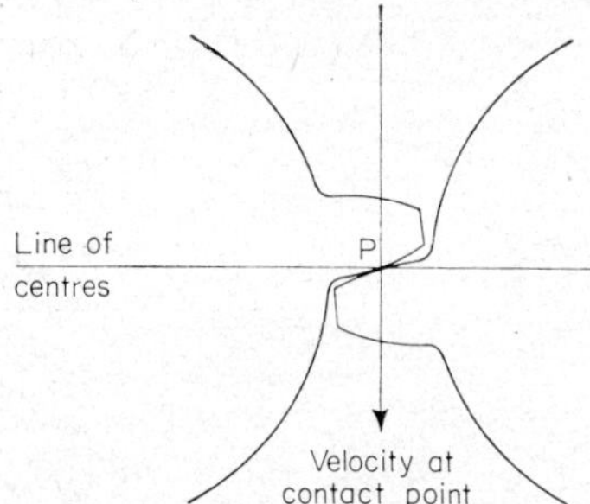

FIG. 3.16. Contact velocities on involute gear.

contact on both surfaces are therefore moving in the same direction at this instant. At all other contacting positions the points of contact of both surfaces are moving in different directions, and therefore some relative sliding takes place between the two surfaces. Because of friction this will result in a loss of energy, and the efficiency of the gears will be less than 100 per cent. The relative, that is to say the sliding, speed, however, is very small compared with the total speed of the gears, and the efficiency of spur gears is high, of the order of 97 per cent. This efficiency is very high compared with that obtained with other forms of rotational drives and gearing, and is a major advantage of spur gears. The efficiency of worm gears, for example, is considerably lower, unless care is taken in their lubrication and design. We shall now consider worm gears briefly, so as to allow for a general discussion on the choice of gear for various textile applications.

3.9 WORM GEARS

Fig. 3.17 shows a part-sectioned view of a worm and wheel gear. As a first approximation the worm resembles a screw, while the wheel can be thought of as a long bolt bent to form an annular ring.

The profile of the teeth cut on to the worm is usually straight-sided, and a line drawn on one face of these teeth at a constant radial distance from the worm axis is a helix. Thus to understand the movements of worm and wheel gears we must first of all describe some of the essential properties of a helix. Fig. 3.18a

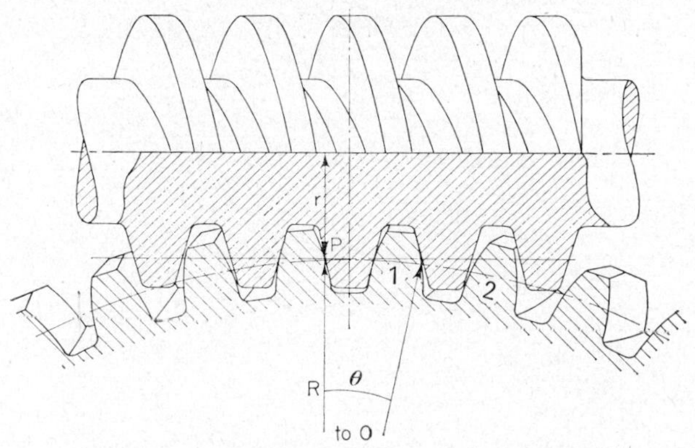

Fig. 3.17. Part-sectioned view of worm gear
(by courtesy of David Brown Corporation (Sales) Ltd).

shows a cylinder of radius r with a helix $APCD$ drawn on it. The essential property of a helix is that at any point on it the tangent to the helix makes a constant angle with the generator of the cylinder drawn through the point being considered.

Thus, in Fig. 3.18a, EPF is a generator of the cylinder (i.e. a line on the surface of the cylinder parallel to its axis) passing through the point P of the helix. PG is the tangent to the helix at P, and this line makes an angle $\frac{1}{2}\pi - \alpha$ with the generator EPF. The angle α, which is constant for all points on the helix, is called the helix angle.

Now suppose we consider a point P moving along the helix from A to D, i.e. through one complete turn of the helix. We want to find out how far parallel to the axis of the cylinder (the helix axis) it has moved, i.e. we want to calculate

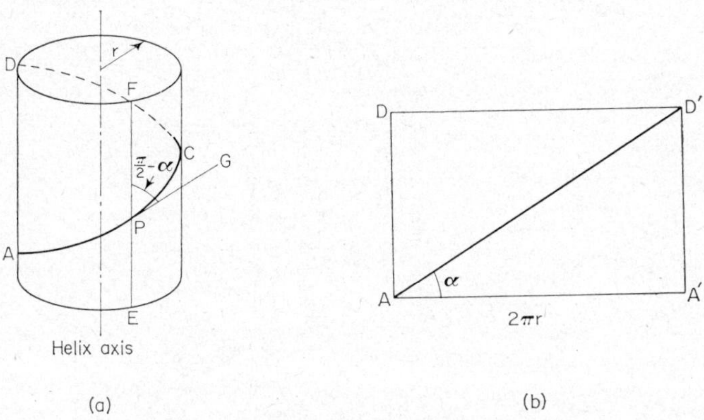

Fig. 3.18. The helix and its properties.

the distance AD. One way of doing this is to imagine the cylinder made of a thin sheet, which can be cut along the generator AD and then opened out into a plane surface. The helix will then appear as a straight diagonal line on the plane, as shown in Fig. 3.18b. In this diagram A, A' and D, D' are coincident points when the rectangle is bent into the cylinder. Thus AA' is equal to the circumference of the cylinder, i.e. $AA' = 2\pi r$. Therefore,

$$AD = A'D' = 2\pi r \tan \alpha$$

This is the result we require for our analysis of the worm gear.

Because the teeth of the worm follow a helical path about the axis of the worm, the teeth on the wheel must be cut at an angle to its axis, and not parallel to the axis as in the case of spur gears. The cross-sectional shape of the teeth, as shown in Fig. 3.17, is normally an involute.

From Fig. 3.17, we see that the point of contact P between the worm and the wheel teeth lies on a helix and, clearly, when the worm has rotated through one complete turn the profile will again be precisely as shown in Fig. 3.17. But, since the worm has pushed tooth 1 forward, this means that tooth 2 will be in the same position as that now occupied by tooth 1 if the worm action is to be continuous. There must therefore be a relationship between the helix angle and the pitch of the wheel teeth. Before considering this relationship it is as well to note that since the wheel moves forward one tooth for each revolution of the worm, the worm will have to make as many revolutions as there are teeth on the wheel in order to rotate the wheel through one complete revolution. In other words, the worm rotates n times faster than the wheel, where n is the number of teeth on the wheel.

To find the helix angle of the worm let us draw a line PO, in the plane containing the worm axis and the wheel centre, perpendicular to the worm axis and passing through the centre O of the wheel. When the contact point P passes through this line as shown in Fig. 3.17, let the distance of P from the worm axis be r, and from the wheel centre be R. These two distances are known as the pitch radii of the worm and the wheel respectively.

Now imagine the worm turned through one revolution. The point of contact P, regarded as a point on the worm, will move forward a distance equal to $2\pi r \tan \alpha$, because it is a point on a helix of radius r and angle α. Regarded as a point on the wheel, P will have moved forward a distance $R \tan \theta$, where θ is the angle subtended by one tooth at the centre of the wheel (see Fig. 3.17), provided that contact is maintained at the same radial distance r (we shall return to this point later). The two distances we have just calculated must clearly be equal if contact is to be maintained, and therefore:

$$2\pi r \tan \alpha = R \tan \theta$$

Now, $\theta = 2\pi/n$ and since θ is a small angle we may write $\tan \theta = \theta$; thus:

$$2\pi r \tan \alpha \simeq R.2\frac{\pi}{n}$$

or

$$\tan \alpha = R/nr \tag{3.3}$$

This equation is approximate, not only because θ is not exactly equal to $\tan \theta$, but because after one rotation of the worm the points on the worm and wheel which were in contact before are no longer in contact. In fact the wheel tooth has moved further out of mesh, and the contact is at a slightly different point. However, since n is a large number and the teeth dimensions are small in comparison with the wheel dimensions the expression for $\tan \alpha$ is reasonably accurate.

We have already noted that the speed ratio of the gear as a whole is given by n so that Equation (3.3) tells us that for large speed reductions α is small, and for small speed reductions α is large. This fact limits the range of speed reductions which can usefully be obtained since the efficiency of the worm gear is greatly dependent on the angle α.

3.10 THE EFFICIENCY OF WORM GEARS

Fig. 3.19a shows the forces acting on the worm, and Fig. 3.19b those on the wheel at a contact point on the central plane of the gear. The reaction Q is at an angle α to the plane at right-angles to the axis of the worm, since the worm surface makes an angle α with this plane at the centre.

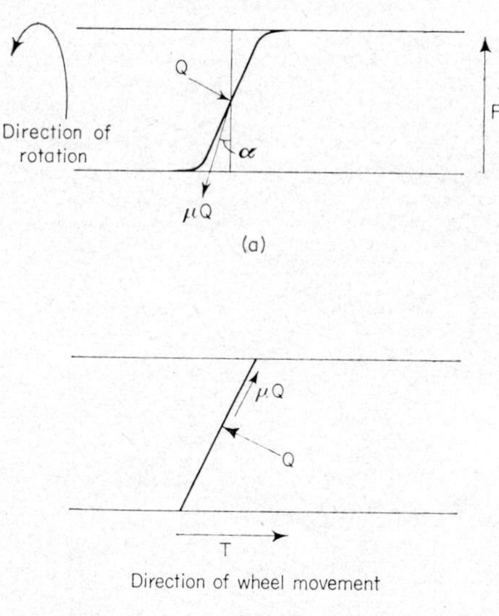

FIG. 3.19. Forces acting on worm and wheel.

The reaction Q is, in fact, the component of the total reaction between the worm and the wheel in a horizontal plane; the other component of the reaction produces a radial load on the wheel and worm bearings, and is of no interest here.

As the worm slides along the face of the tooth equal and opposite frictional forces μQ will be set up as shown. If F is the circumferential component of

From Equation (3.6) it can be seen that under such circumstances the speed ratio of the train will be directly proportional to the number of teeth on gear D (as it is a driven gear of the gear pair CD). Hence it is possible to derive the draft obtained directly from a knowledge of the number of teeth on gear D. In fact in the train shown in Fig. 3.22 the draft obtained is half the number of teeth on gear D. The draft ratio can therefore be altered only in steps of a half since, for example, a 40-tooth gear will give a draft ratio of 20, while if gear D is changed to the next larger gear, namely a 41-tooth gear, the draft will be altered by half a unit to a ratio of $20\frac{1}{2}$. It can therefore be seen that the use of idler gears can provide a more accurate or a more convenient solution to the problem of obtaining exact gear ratios.

It should be noticed that even if idler gears are used it is only possible to obtain ratios which are expressible as a fraction, and as the number of teeth on any gear should not normally exceed 100 the actual ratios that can be obtained are always in relatively large steps. It is, however, always possible to obtain a ratio which is correct to within 0·5 per cent by using a pair of gears, provided there is no restriction on the centre distance of the two gears. This is sufficiently accurate for most textile applications. Various methods have been devised for finding the gear ratio which gives the closest approximation to a required speed ratio. For details of these the reader is referred to the work by Buckingham, given in the bibliography at the end of this chapter.

3.12 EPICYCLIC OR PLANETARY GEAR TRAINS

Fig. 3.23 shows an example of an epicyclic, or planetary, gear train. A feature of such gears is that they contain at least one gear wheel that is mounted on an arm, the arm being free to rotate about a fixed point. In the example shown, gear 2 is mounted on arm 3, which can rotate about the fixed point F. Gear 2

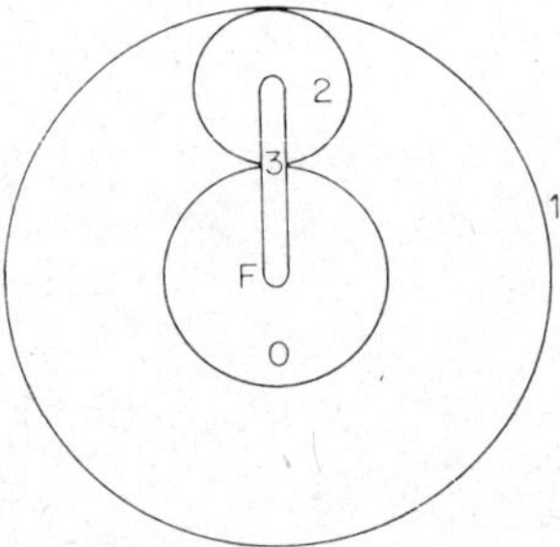

Fig. 3.23. Planetary gear train.

is in mesh with two other gears, 0 and 1, which have their centres at F. When the gear train is in operation, gear 2 rotates about its own axis while this axis, being fixed in arm 3, rotates about F. This motion is similar to that of a planet about the sun; hence gear 0 is known as the sun gear, and gear 2 is called the planetary gear.

It is not at all obvious what the relative movements of the various parts of this gear will be, but the following artifice makes it relatively simple to determine such motions. We shall suppose that gear o has n_0 teeth, gear 1 has n_1 teeth and gear 2 has n_2 teeth. Now imagine that the arm 3 is held fixed in space, while gear 1 is made to turn through one complete revolution. Since the arm is fixed, the system of gears can be regarded as a simple gear train, like that

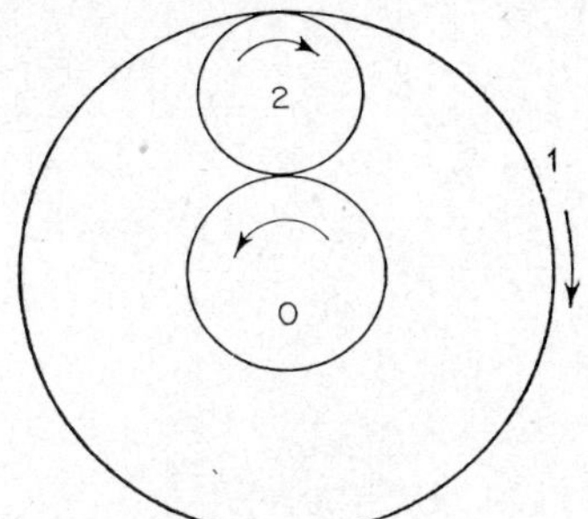

FIG. 3.24. Simple analogous gear train.

shown in Fig. 3.24. The directions of rotation of each gear have been indicated in this diagram. Note that gears 1 and 2 rotate in the same direction while gear o rotates in the opposite direction.

Consider first what happens to gear 2. As gear 1 makes one complete revolution its n_1 teeth mesh with the teeth on gear 2, hence gear 2 must make n_1/n_2 revolutions. Similarly gear o makes n_2/n_0 revolutions (in the opposite direction) while gear 2 makes one revolution. But gear 2 makes n_1/n_2 revolutions; hence gear o must make $(n_1/n_2) \times (n_2/n_0) = n_1/n_0$ revolutions, while gear 1 makes one revolution. These relative motions have been entered in the first line of Table 3.2, the minus sign under gear o indicating that this gear is rotating in a direction opposite to that of gear 1.

What we have found so far are the motions of the individual gear wheels *relative to the arm 3*, which has been kept fixed in space. If we now lock the gears relative to the arm, and allow the arm to make N complete revolutions, then each individual gear wheel will also make N revolutions about the fixed point F. This motion is shown in the second line of Table 3.2.

The total relative motion of the gears is then found by adding together the two separate relative motions we have just discussed, and these sums are given in the bottom line of Table 3.2.

Table 3.2
Movement of Members of a Simple Epicyclic Gear Train

Condition Imposed on Gear	Gear 1	Gear 2	Arm 3	Gear o
Gear 1 moved through one turn, arm fixed	1	$+\dfrac{n_1}{n_2}$	o	$-\dfrac{n_1}{n_0}$
Gears locked to arm, which rotate through N revolutions	N	N	N	N
Total relative movement	$1 + N$	$N + \dfrac{n_1}{n_2}$	N	$N - \dfrac{n_1}{n_0}$

From this table we see that, if ω_0, ω_1, ω_2, ω_3 are the angular velocities of the gear wheels and arm, then, for example,

$$\frac{\omega_0}{\omega_1} = \frac{N - (n_1/n_0)}{1 + N}; \quad \frac{\omega_3}{\omega_2} = \frac{N}{N + (n_1/n_2)}$$

and so on, where N is a number determined by certain additional restraints imposed on the gear train.

For instance, a simple use of this gear is to hold gear 0 fixed in space. The arm is used as the driving member of the gear train, and gear 1 is the driven member. Since gear 0 is fixed in space, we must have $\omega_0 = 0$. But from Table 3.2

$$\frac{\omega_0}{\omega_3} = \frac{N - (n_1/n_0)}{N}$$

so that N must equal n_1/n_0 if ω_0 is to be equal to zero. Therefore the ratio of the speeds of the driven and driving members is:

$$\frac{\omega_1}{\omega_3} = \frac{1 + N}{N} = \frac{1 + (n_1/n_0)}{n_1/n_0} = \frac{n_0 + n_1}{n_1}$$

Another use of this particular gear is to drive both gear 0 and the arm at speeds ω_0 and ω_3 respectively. The speed ω_1 of the driven gear 1 is therefore a function of both ω_0 and ω_3. From Table 3.2 we see that:

$$\frac{\omega_1}{\omega_3} = \frac{1 + N}{N}$$

and

$$\frac{\omega_0}{\omega_3} = \frac{N - (n_1/n_0)}{N}$$

Eliminating N we find that:

$$\omega_1 = \frac{\omega_3(1 + n_1/n_0) - \omega_0}{n_1/n_0}$$

Such a gear can therefore be used to drive a shaft at a speed which is the difference between two speeds. These simple examples illustrate the main purposes of epicyclic gearing: (a) to act as an ordinary gear train—it will be shown later, by considering further examples, that certain epicyclic gear trains can have very high speed reduction ratios—and (b) to use the gear to add or subtract angular velocities.

To clarify these two applications it is necessary to consider two more practical, but more complex, forms of epicyclic gearing. Fig. 3.25 shows an epicyclic gear train. Gear 0 is a fixed gear on which a planetary gear 2 is rolling. Gear 2 is driven by means of the internal driving gear 1. The shaft of gear 2 is attached to a carrier arm 3, so that while it rotates about its own subsidiary shaft, this subsidiary shaft itself rotates about the main shaft which passes through the centre of the fixed gear 0. To the other end of the subsidiary shaft is attached gear 4, which drives the output gear 5. The speed reduction of this gear is the

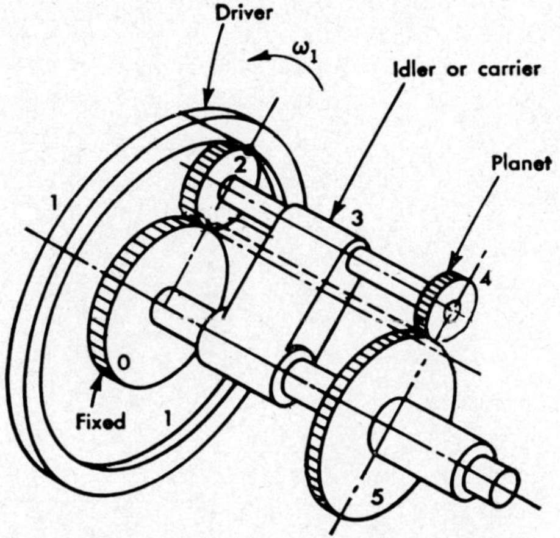

FIG. 3.25. Planetary gear train.

ratio of the angular velocity of gear 1 to that of gear 5, and to find this gear ratio
in terms of the number of teeth on the various gear wheels we proceed as before.
First of all, the arm 3 is fixed in space, and the system then becomes equivalent
to an ordinary (though fairly complex) gear train.

If gear 0 is now turned through one complete revolution, the other gears
will rotate through the number of revolutions given by the expressions shown
in the first line of Table 3.3, where n_0, n_1, n_2, n_4 and n_5 are the number of teeth
on gears 0, 1, 2, 4 and 5 respectively.

Table 3.3

Condition Imposed on Gear	Gear 0	Gear 1	Gear 2	Arm 3	Gear 4	Gear 5
Fix arm 3 and rotate gear 0 through one revolution	1	$-\dfrac{n_1}{n_0}$	$-\dfrac{n_0}{n_2}$	0	$-\dfrac{n_0}{n_2}$	$\dfrac{n_0}{n_2}\cdot\dfrac{n_4}{n_5}$
Lock gears relative to arm 3 and rotate arm through N revolutions	N	N	N	N	N	N
Total relative movement	$1+N$	$N-\dfrac{n_0}{n_1}$	$N-\dfrac{n_0}{n_2}$	N	$N-\dfrac{n_0}{n_2}$	$N+\dfrac{n_0}{n_2}\cdot\dfrac{n_4}{n_5}$

Now we lock the gears relative to arm 3, and rotate the arm through N
complete revolutions, as shown in Table 3.3. The total relative movements
are given in the last line of this table. However, we know that gear 0 is fixed;
hence we must have

$$1 + N = 0$$

or

$$N = -1$$

We want to find the ratio ω_1/ω_5 which, from Table 3.3, is given by

$$\frac{\omega_1}{\omega_5} = \frac{1 + \dfrac{n_0}{n_1}}{1 - \dfrac{n_0 n_4}{n_2 n_5}}$$

when $N = -1$. This expression can be simplified somewhat, since it can be seen that since gear 0 and gear 2 fit into gear 1, we must have

$$\frac{d_0}{2} + d_2 = \frac{d_1}{2}$$

or

$$n_0 + 2n_2 = n_1$$

Further, since the distance between the centres of gears 4 and 5 is the same as the distance between the centres of gears 0 and 2 (they use the same shafts), we must have

$$n_0 + n_2 = n_4 + n_5$$

Eliminating n_0, we find that the speed reduction becomes

$$\frac{\omega_1}{\omega_5} = \frac{2n_2 n_5}{n_1(n_2 -- n_4)}$$

Since n_2 can be made very nearly equal to n_4 it is possible by this means to obtain very large speed reduction ratios. As an example, let us consider the case where $n_1 = 200$, $n_2 = 50$, $n_4 = 49$, and $n_5 = 101$. The gear reduction ratio is then $50 \cdot 5 : 1$. With other forms of epicyclic gearing even higher reduction ratios can be achieved, using only four gears. To consider the other type of epicyclic gear which is used for adding or subtracting speeds we must turn to a completely different type of gear train, the differential gear.

3.13 THE DIFFERENTIAL GEAR

Fig. 3.26 shows a typical differential gear. The two bevel gears A and B are mounted on shafts fixed in a housing that can itself rotate about the axis of the other two meshing bevel gears D and E. This casing is driven through gear C at one of the input speeds. The third bevel D is driven at the second input speed and the fourth bevel E drives the output shaft.

We shall now investigate the relation between the angular speed ω_e of the driven gear E and the input speeds ω_c and ω_d of the driving gears C and D. The system can be analysed using the same method as before. We need consider only gears C (the casing), D and E. The reason for this is that D and E, being of the same diameter, must have the same number of teeth, and the intermediate gears A and B act simply as idler wheels.

If we fix the casing in space, and rotate gear D through one revolution, then gear E will rotate through one revolution in the opposite direction. This motion is shown in the first line of Table 3.4. The gears are now locked and the whole

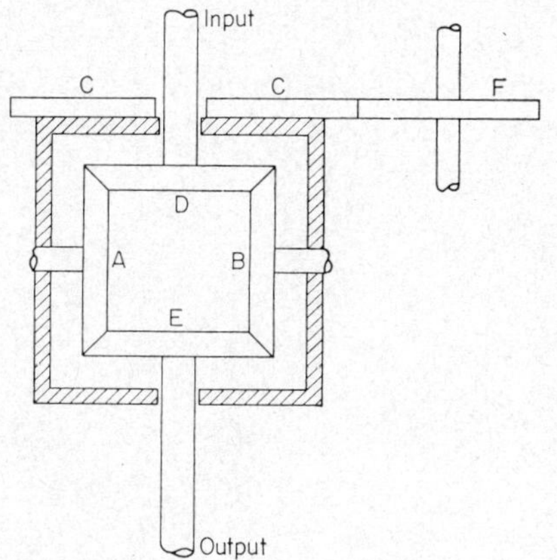

FIG. 3.26. Differential gear.

system is rotated through N revolutions. We are thus led to the relative movements shown in the last line of Table 3.4.

Table 3.4
Gear movements in Differential Gear

Condition Imposed on Gear	Casing C	Gear D	Gear E
Fix casing C and rotate D through one revolution	0	1	−1
Lock gears and rotate whole system through N revolutions	N	N	N
Total relative movement	N	$N + 1$	$N − 1$

We therefore have

$$\frac{\omega_d}{\omega_c} = \frac{N + 1}{N}$$

and

$$\frac{\omega_e}{\omega_c} = \frac{N - 1}{N}$$

Eliminating N, we obtain:

$$\omega_e = 2\omega_c - \omega_d$$

The angular speed of the driven gear E is thus equal to the difference between twice the driving speed of casing C and the driving speed of gear D.

If gear C is itself driven through a straight spur gear F (as shown in Fig. 3.26) having half the teeth of gear C, then

$$\frac{\omega_c}{\omega_f} = -\frac{1}{2}$$

the minus sign having been introduced since gears F and C rotate in opposite directions. The speed of gear E is then given by

$$\omega_e = -(\omega_f + \omega_d)$$

i.e. the speed of gear E is equal to the sum of the speeds of gears F and D.

The differential gear is widely used to control the speeds of various let-off and take-up devices. In these mechanisms it is necessary to maintain a constant surface speed of the yarn packages such as the warp beam of a warp knitting machine, or the bobbin on a cone-flyer package. Since the diameter of the package varies as the bobbin fills up, or the warp becomes empty, it is necessary for the rotational speed to be altered. This could be done by driving the beam or bobbin directly from a varying speed device which would vary in such a way as to compensate for the change in package diameter. Such a device is perfectly feasible for warp knitting machines, and is in fact often used. However, for a flyer-bobbin package it should be noted that the speed of the package must be the difference between the flyer speed and the speed needed to wind onto the package.

The range in winding-on speeds is dictated by the change in bobbin diameter, and the latter will remain within fixed limits. The flyer speed, however, must be capable of being changed to produce variations in twist rate when this is required. The bobbin speed must therefore be variable in such a way that the speed of rotation of the empty bobbin can be changed, but the *difference* between the speed of the empty and full bobbin must always be the same. It would require very complex changes to be made to obtain such an alteration in the speed range from a single variable speed drive, and this is completely obviated by using a differential gear as shown in Fig. 3.27.

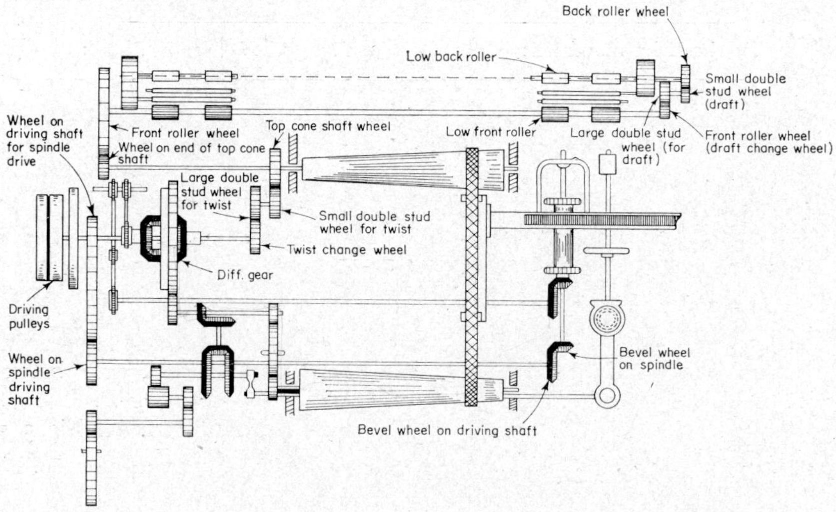

FIG. 3.27. Gear train for cone-flyer variable-speed drive.

The main drive to the differential comes from the same drive as the drive to the flyer, while the casing (or cage) of the differential is driven from the variable speed cones. The output from the differential drives the bobbin. If the cage is stationary the bobbin will run at the same speed as the flyer; as the drive from the variable speed device is brought into operation the two are

FIG. 3.28. Epicyclic gear for sliver feed
(by courtesy of Carding Specialists Ltd).

driven at a speed difference directly proportional to the speed of the variable-speed device. Hence the variable-speed drive can be made directly dependent on the winding-on speed and does not need to be altered when the twist level is changed.

3.14 FURTHER USES OF EPICYCLIC GEAR TRAINS

We have so far considered gear trains in which the final drive is a simple rotation. It is possible, however, to use planetary gears to obtain a rotation about a rotating centre. Fig. 3.28 shows an epicyclic gear used to feed sliver into a can. To obtain a satisfactory package in large cans, so that the sliver can be readily removed, it is desirable that the sliver should be fed in to the can in a series of overlapping coils, as shown in Fig. 3.29. To obtain such a feed the sliver is fed

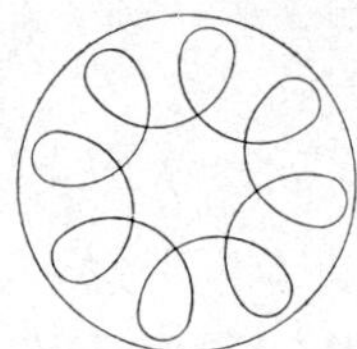

FIG. 3.29. Required feeding motion.

into a tube H which rotates with the planetary gear marked B in Fig. 3.28. This gear rotates about its own axis, and this axis, which is secured to a plate A that acts as the arm of the planetary gear, also rotates about the centre of the whole gear train, so laying down the sliver in the required fashion. The gear C is merely used to drive the feed rollers supplying the sliver to the tube H.

A similar application using bevel gears is shown in Fig. 3.30, which shows

FIG. 3.30. Front draft-twist rollers from woollen drafting frame.

the upper and lower parts of the gear drive to the twist rollers of a woollen spinning frame. These rollers are the front rollers of the frame, and are required to feed drafted material forward while the rollers rotate about the sliver axis to insert false twist into the drafting zone. To accomplish this the two rollers are driven by the cage bevels of a differential gear, so that the rotational speed of the bevels about their own axes produces the forward movement of the sliver, and the rotation of the cage inserts the false twist. The two cage bevels have been replaced by two skew gears, which are very similar in action to the bevels, so as to enable the axis of the two rollers and their driving gear to be displaced from the centre line of the cage, since the rollers themselves must meet on the centre line.

3.15 THE EFFICIENCY AND CHOICE OF GEAR TRAINS

We have so far considered various gear types, of which spur gears have the highest efficiency. Simple spur gear trains also have high efficiency since, if the efficiency on a pair of gears in a train is X, the efficiency of two pairs of gears in a simple gear train is X^2; for three pairs of gears it is X^3, and so on. As X is of the order of 97 per cent the efficiency of simple spur gear trains is always high. The same cannot be said of epicyclic gears, which can have relatively poor efficiencies. Simple gear trains are therefore the first choice for any gear train where special requirements do not have to be met. These other requirements can sometimes be simple, as for example where it is necessary to change the direction of rotation through a right-angle, when a worm gear may be preferred, but they may also be complex, as the following example shows.

The drive to the take-up of a loom requires a very large speed reduction, because the take-up roller rotates about once in 1 to 10 minutes, depending on the type of cloth woven. A gear train consisting of 7 gears was used to obtain

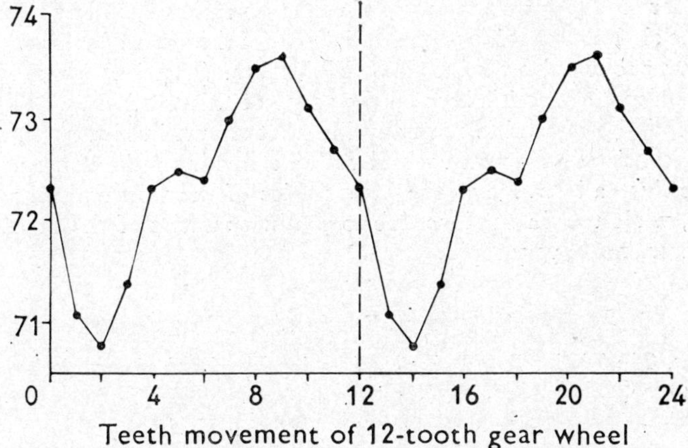

FIG. 3.31. Variation in speed of take-up roller
(by courtesy of Wool Industries Research Association).

the required speed reduction, but it was found that the cloth produced had periodic marks on it due to the uneven movement of the take-up roller. In-accuracies in the gear teeth can give rise to such uneven movements, but they can also arise due to the gears being eccentrically mounted so that the pitch diameter varies as the gear rotates. The first cause produces a variation every time a tooth meshes, while the second produces a variation each time a gear completes a full turn. Fig. 3.31 shows the measured variation in speed of the take-up roller due to a very badly worn gear. Such errors in the teeth of the earlier gears in the train produce relatively high frequency variations in the roller since these gears are rotating rapidly, while faults in the later gears produce a slow variation in speed. While the use of accurately cut and mounted gears

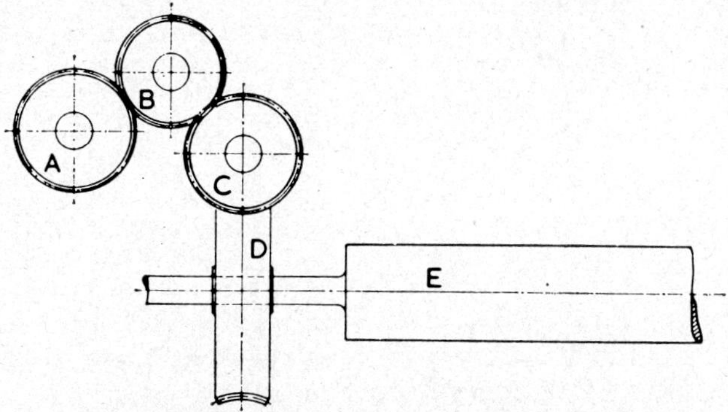

FIG. 3.32. Preferred gear design for the take-up roller.

considerably improved the quality of the product it did not eliminate the fault completely. It was found, however, that relatively short frequency variations (of the order of $\frac{1}{8}$ inch movement of the take-up roller) did not produce visible flaws in the cloth and that long term variations (of the order of 10 inch movement of the take-up roller) also produced no visible flaws. (This is due to the behaviour of the let-off motion, and is beyond the scope of this text.) The gear train was therefore redesigned, and is normally like that shown in Fig. 3.32, where the final stage consisted of a worm wheel with a reduction ratio of 150 : 1. By this means the fluctuations due to the inaccuracies of the worm wheel were made very slow, while the fluctuations due to the inaccuracies of the earlier teeth in the train were made very rapid.

Suggested Further Reading

R. Y. CASE. *The Timing Belt Drive Engineering Handbook*. New York Belting and Packing Co, Philadelphia, 1954.
J. S. BEGGS. *Mechanism*, chapter 3. McGraw-Hill, 1955.
E. BUCKINGHAM. *Spur Gears*. McGraw-Hill, 1928.

CHAPTER 4

CAMS

4.0 INTRODUCTION

All the mechanisms considered in the previous chapter have been concerned with the transference and alteration of a rotational motion (usually provided by an electric motor drive) from one part of a machine to another and, as we have seen, the final motion is in some cases very complex. In this chapter we shall be concerned with mechanisms that are used to convert rotational motion into a to-and-fro motion. An important example of such a motion is the to-and-fro movement of a shuttle or reed in a loom.

It is relatively simple to convert a rotational motion into a straight-line movement in one direction. For example, a gear meshing with a rack—i.e. a gear cut onto a flat bar which can be considered to be a gear of infinite radius—will move the rack forward as long as the gear rotates. Such a mechanism can provide a to-and-fro motion when the rotating shaft is oscillating, but methods of obtaining such an oscillation have not yet been considered and, as will be seen later, involve the use of a to-and-fro motion at an earlier stage in the whole mechanism. We must therefore consider, first, methods of obtaining a to-and-fro motion from a continuous rotation.

To obtain a to-and-fro motion from a continuous rotation it is necessary to use either a cam or a linkage mechanism. Such mechanisms are in many ways complementary and can often be used as alternative methods of obtaining the same or similar final motions. The cam mechanism in all its various forms is a vital part of most textile machines because of its versatility, and will consequently be considered first. Linkage mechanisms are considered in the next chapter.

4.1 CAMS

Fig. 4.1 shows a simple cam system. The large disc A, which has a known shape, is in contact with the knife edge B at the lower end of the shaft C. The shaft is constrained to move in an up-and-down fashion, the constraint being provided by the slot D. The knife edge and shaft are known as the cam follower. As the cam disc is rotated in the direction shown in the diagram, it can be seen that, since the radial distance between the centre of rotation O of the cam and the point of contact B, where it rubs against the follower, is increasing as the disc rotates, the cam disc will compel the follower to move upwards. As the disc rotates further a point will be reached where the radial size of the cam begins to decrease. The follower will then move downwards, either under its own weight or because of the action of a spring, so that it remains in contact with the cam, and will

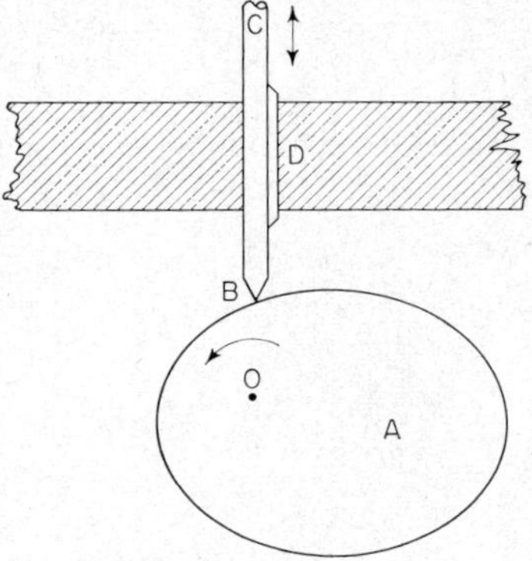

FIG. 4.1. Disc cam with knife-edged follower.

continue to move down as the cam rotates further. Clearly, such a cam can produce a to-and-fro motion from a steady rotation.

Cams fall into various classes depending on the kind of cam disc used and the type of follower. The cam may consist of a disc as shown in Fig. 4.1, or of the cut-away portion of a cylinder as shown in Fig. 4.2. The latter, known as a

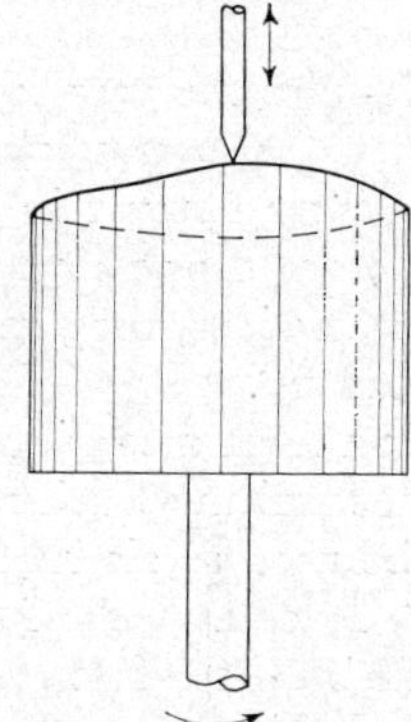

FIG. 4.2. End cam.

cylindrical cam, moves the follower in a direction which is parallel to the axis of rotation of the cam, while the disc cam moves the follower in a direction at right-angles to the axis of rotation of the cam. Fig. 4.3 illustrates the three kinds of followers that are commonly used. Fig. 4.3a shows a knitting needle being operated on by a cylindrical cam. This is a typical example of a knife-edged follower. Fig. 4.3b shows a mushroom follower driven by a disc cam, and Fig. 4.3c shows a roller follower. The last type of follower, since it interposes a roller between the follower and the cam, eliminates sliding between the follower and cam, and so minimises wear between the parts. It is consequently the most common type of follower used. Fig. 4.4 shows the main cam shaft of a flat-bed knitting machine which controls most of the movements required for knitting a fully fashioned stocking. It can be seen that it consists of some forty disc cams with roller followers.

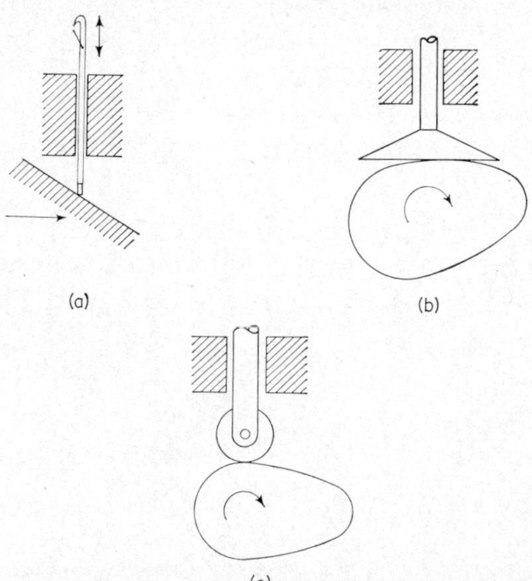

FIG. 4.3. (a) Knitting cam; (b) mushroom follower and disc cam; (c) roller follower.

FIG. 4.4. Cams used to control movements of the parts of a flat-bed knitting machine.

The cams described above are known as negative cams since, while the follower is pushed positively by the cam in one direction, it returns under the force of its own weight or that of a spring. We shall see later that cams can very easily produce unwanted vibrations in the movement of the follower, and the size of such vibrations can be very great when negative cams are used. To minimise the amplitude of the vibration the cam can be made positive in action. Positive cams consist essentially of two cams acting on one follower; one to drive the follower upwards, and one to return it.

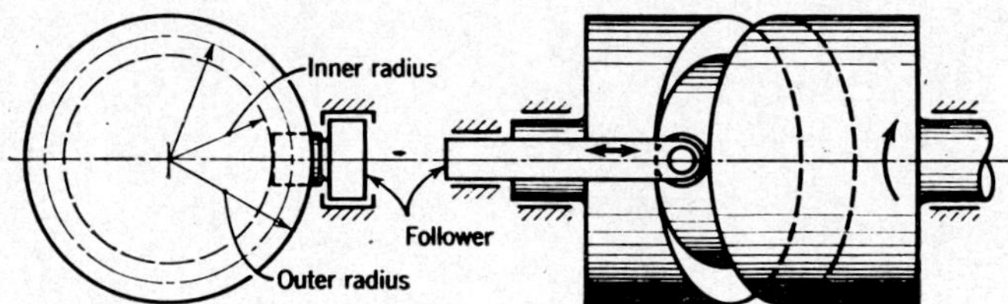

FIG. 4.5. Cylindrical cam.

If the cam is cylindrical this presents no problem, since a positive cylindrical cam can be made by cutting a groove in a cylinder, as shown in Fig. 4.5. The lower edge of the groove drives the follower upwards, while the upper edge drives it downwards. It should be noted that the cams used on a weft knitting machine are positive cylindrical cams, since the return upward motion is accomplished by the direct action of the lower surface of the cams.

Fig. 4.6 shows a positive cam system using disc cams and roller followers. The disc has a groove cut into it, in which the roller of the follower rests. The roller is driven alternately from one side or the other of the groove according to whether the follower is being raised or lowered. This type of positive cam has

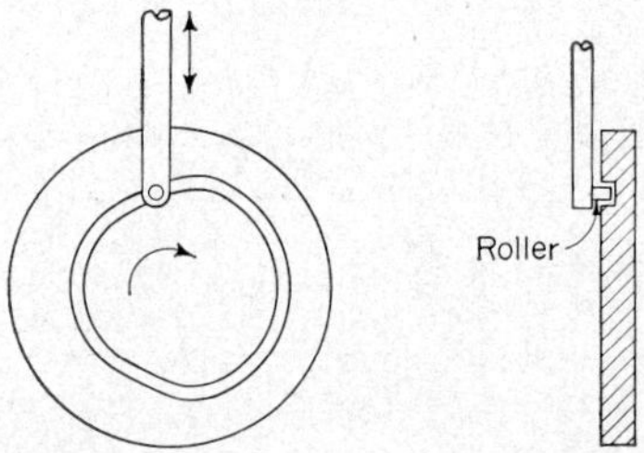

FIG. 4.6. Face groove cam.

certain disadvantages which can be overcome by the use of the system shown in Fig. 4.7. In this system the follower has been forked and has two rollers attached to it. The right-hand roller is in contact with disc 1 while the left-hand roller is in contact with disc 2, which lies behind disc 1. Both discs are driven from the same shaft and rotate together. They do, however, have different shapes so that when a movement in one direction is required the right-hand follower is in contact with cam disc 1, and is propelled in that direction. When the return motion is required the cam disc 2 comes into contact with the left-hand roller, and returns the follower. Cam disc 1 must be cut away at this point so that the right-hand follower does not jam when disc 2 is driving the follower, and vice versa. Such positive cam systems can be seen in older models of warp knitting machines, where they were used to drive the knitting needle bars.

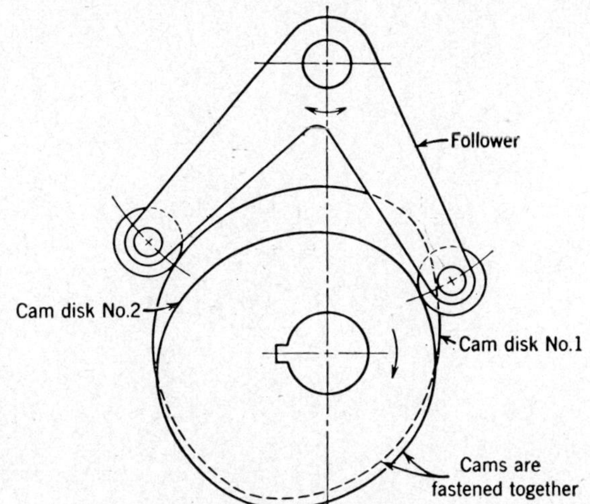

FIG. 4.7. Conjugate cams.
(Reproduced from H. A. Rothbart, *Cams-Design, Dynamics and Accuracy*, Wiley, 1956.)

4.2 THE MOTION OF THE CAM FOLLOWER

The exact nature of the to-and-fro motion of a follower is dependent on the exact shape of the disc, and can be very complex if necessary. In particular, it should be noted that if it is required that the follower should be stationary at any point of the cycle, it is only necessary to see that the cam shape at this part of the cycle is of constant radius; in other words, the cam over this part of the cycle should be a section of a circle. Such a halting in the movement of the follower is known as a dwell. By varying the size of the dwell it is possible to compress the to-and-fro movement into any required part of the cycle of the whole machine action.

To demonstrate how it is possible to find the follower motion, given a known cam shape and speed, let us consider the simple example shown diagrammatically in Fig. 4.8. The cam disc is a circle of radius r mounted eccentrically a distance a from the centre of the circle. The mushroom follower is constrained to move along a line at right-angles to the face of the follower. At the point of rotation shown in the diagram the centre of the disc C must be a distance a from the centre of rotation R, and CR makes an angle θ with a fixed line RQ drawn through R at right-angles to the direction along which the follower moves. The variation in the distance from this fixed line to the mushroom follower face is clearly the variation in position of the follower. The line CA drawn from the centre C of the disc to the point of contact A between follower and disc must be at right-angles to the face of the mushroom follower, since the follower face must be a

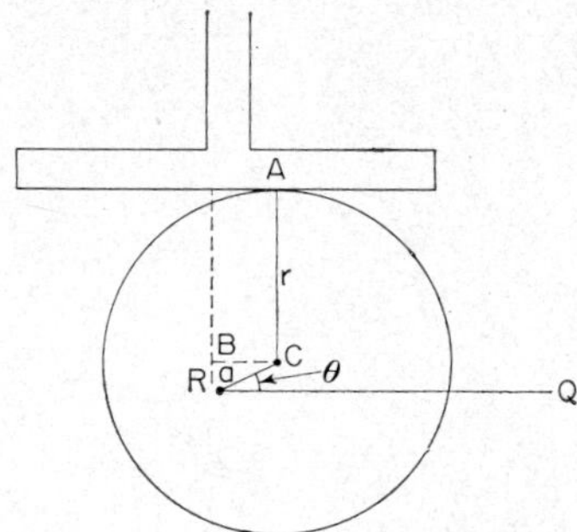

FIG. 4.8. Cam which will produce
a simple harmonic motion.

tangent to the surface at point A. It follows therefore that the distance from the fixed line QR to the mushroom face is (see Fig. 4.8)

$$AC + BR = r + a \sin \theta$$

As the disc is rotated, a and r remain constant and θ changes. Clearly the movement x of the follower can be represented at any rotational position θ of the disc by

$$x = a \sin \theta \qquad (4.1)$$

where x is the height of the follower above the position it had when $\theta = 0$, i.e. when RC coincided with RQ.

If the cam is rotating at a constant rotational speed ω, then, clearly, $\theta = \omega t$ where t represents time measured from the instant when θ was zero. The movement of the mushroom follower is therefore given by

$$x = a \sin \omega t \qquad (4.2)$$

This equation is the well-known equation for a simple harmonic motion, and hence the cam system shown in Fig. 4.8 produces a simple harmonic motion in the follower.

4.3 THE LINEAR CAM

In practice, the process described above is usually reversed, i.e. it is necessary to find the cam shape that will produce a desired motion in the follower. To illustrate this problem, let us consider the case of a disc cam with a knife-edged follower which is required to move the follower upwards with a constant velocity, and immediately the follower reaches its maximum height it is returned at the same constant velocity to its original position, when the cycle is recommenced. The minimum radius of the disc is to be b.

Fig. 4.9 shows the cam and follower at an instant during the upward motion of the follower. R is the centre of rotation of the cam, A is the point of contact of the cam and the follower, and RB is the minimum radius of the cam. Obviously the follower will be at the lowest point of its motion when A coincides with B, i.e. when RB is vertical. We measure the time t from this instant, and consider the cam when RB has turned through an angle θ from the vertical.

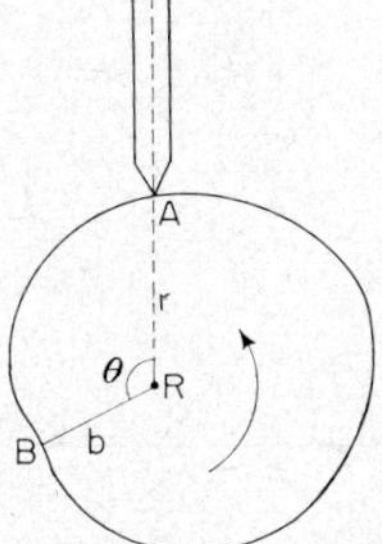

Fig. 4.9. Linear or heart-shaped cam.

If we denote the height of the follower above its lowest point by x, then evidently

$$x = r - b \qquad (4.3)$$

where r is the radial distance RA in the position considered.

Now if the follower is to move upwards with constant velocity k, we must have

$$x = kt$$

so that Equation (4.3) may be written:

$$r = b + x = b + kt \qquad (4.4)$$

Suppose now that the maximum height of the follower above its lowest position is $x = a$. Then:

$$a = kt_m \qquad (4.5)$$

where t_m is the time taken to move from the lowest to the highest position. Also, since the follower is to return to its lowest point at the same constant velocity, it must reach its highest point when $\theta = \pi$. Therefore, if the cam rotates with constant angular speed ω, we must have

$$\theta = \omega t$$

and

$$\pi = \omega t_m \qquad (4.6)$$

Eliminating t_m from (4.5) and (4.6):

$$k = a\frac{\omega}{\pi}$$

Therefore Equation (4.4) becomes:

$$r = b + \frac{a\omega}{\pi}\left(\frac{\theta}{\omega}\right) = b + \frac{a\theta}{\pi} \qquad (4.7)$$

which is the polar equation of the cam outline for $\theta \leqslant \pi$, i.e. when the follower is rising.

G

When $\theta > \pi$, i.e. when the follower is descending, the height x is given by

$$x = a - k(t - t_m)$$
$$= a - kt + kt_m$$
$$= a - \frac{a\omega}{\pi} \cdot \frac{\theta}{\omega} + \frac{a\omega}{\pi} \cdot \frac{\pi}{\omega}$$
$$= 2a - \frac{a\theta}{\pi}$$

so that for $\theta \geqslant \pi$

$$r = b + x = b + a\left(\frac{2\pi - \theta}{\pi}\right) \tag{4.8}$$

Equations (4.7) and (4.8) give us the required shape of the cam disc in terms of the radius r and the angle θ. This shape can be found graphically in a very simple way by choosing a set of values of θ (usually $\pi/6$, $\pi/3$, $\pi/2$, etc.—that is 30°, 60°, 90°, etc.), and then by direct calculation finding the value of r at each of these angles. The shape of the cam can then be plotted. This process has been carried out on Fig. 4.9, and the shape shown is the required shape of the cam (the well-known heart-shaped cam).

This process can clearly be carried out in the same way for any required movement, for which x is known as a function of time. The above example

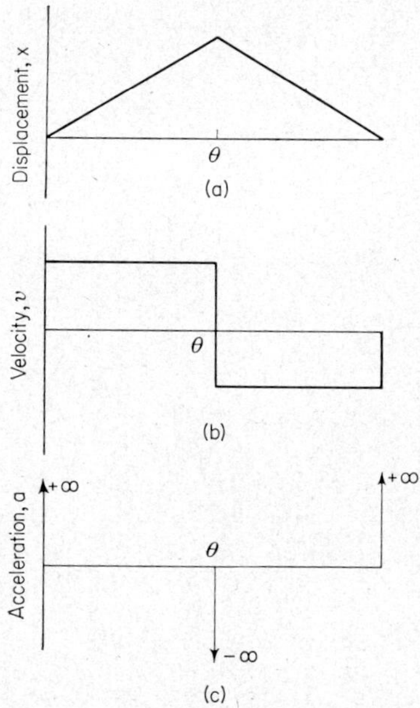

FIG. 4.10. Displacement, velocity and acceleration of the follower
with a linear cam.

illustrates the fact that it is not always necessary to know the rotational speed of the cam (so long as it is constant), since the requirements can usually be written in the form of an equation relating x to θ, instead of t. Because of this the profile of any cam can be found once the relationship between x and θ is known, and it is usual to discuss the various cam shapes in terms of the graphs of x versus θ, rather than in terms of the profile of the actual cam used. Thus Fig. 4.10a shows the graph of x versus θ for the cam previously considered. It is because of the shape of this graph that such a cam is usually known as a linear cam.

This method of representation is particularly useful because it can be used to find the speed and acceleration of the follower very readily. To do this we make use of the property of the multiplication of derivatives thus:

$$\frac{dx}{dt} = \frac{dx}{d\theta} \cdot \frac{d\theta}{dt}$$

where the velocity, v, of the follower is by definition dx/dt, and similarly $d\theta/dt$ is the rotational speed of the cam—usually constant and equal to ω. $dx/d\theta$ is the slope of the graph of x versus θ and can readily be found. Hence:

$$v = \omega \frac{dx}{d\theta} \tag{4.9}$$

In addition d^2x/dt^2 is the acceleration, a, of the cam follower and can be written as follows:

$$\frac{d^2x}{dt^2} = \frac{d}{dt}\left(\frac{dx}{dt}\right) = \frac{d}{d\theta}\left(\frac{dx}{dt}\right)\frac{d\theta}{dt} = \omega\frac{d}{d\theta}\left(\frac{dx}{dt}\right)$$

But

$$\frac{dx}{dt} = \omega\frac{dx}{d\theta}$$

and hence, since ω is a constant

$$a = \omega\frac{d}{d\theta}\left(\omega\frac{dx}{d\theta}\right) = \omega^2\frac{d^2x}{d\theta^2} \tag{4.10}$$

$d^2x/d\theta^2$ is known from the relationship between x and θ, and thus it is possible to find the acceleration of the follower.

Let us use Equations (4.9) and (4.10) to discover the velocity and acceleration of the cam follower motion shown in Fig. 4.10a. The value of $dx/d\theta$ is clearly a constant during the rise and during the fall section, but since the direction of motion of the follower is opposite in the two cases, the motion consists of a constant positive velocity followed by a constant negative velocity as shown in Fig. 4.10b. Except at the beginning, middle and end of the motion the acceleration or rate of change of velocity is zero since the velocity is constant. At the start and end of the cycle the velocity is changed suddenly from a negative value to a positive value in an infinitely short time. In other words, the rate of change of velocity is positive and infinitely large. At the centre of the cycle the velocity is changed suddenly from a positive to a negative value, i.e. the acceleration is now negative and infinitely large. Fig. 4.10c shows the acceleration at various times of the cycle.

The existence of these infinitely large accelerations poses a problem, since we must remember that the follower has a mass and, according to Newton's laws of motion, the force on the follower needed to overcome its inertia is the product of its mass times its acceleration. An infinite acceleration therefore requires an infinite force to produce it, which is clearly not possible. In fact, heart-shaped or linear cams do operate fairly successfully in practice, because the material at either the cam and follower surfaces or in the follower itself is distorted by the force so that the actual motion deviates from the theoretical values, to produce large but non-infinite accelerations at the start and centre of the cycle. The effect of elasticity will be considered later in this chapter, but at this point let us consider whether it is possible to alter the required motion to prevent these sudden large accelerations from occurring.

4.4　THE PARABOLIC CAM

There are a few examples of textile mechanisms in which the shape of the cam is completely determined by the requirements of the machine. For example, a lifting mechanism for the ring rail of a ring spinning machine must move up and down at a constant velocity. If it does not do so, the package produced will not be straight-sided and will therefore not be usable. In such circumstances the whole of the cam profile required is completely determined, and no alterations can be made to it to prevent the sudden accelerations and decelerations that such a movement will inevitably produce. There are many cases, however, where the beginning and end of a motion is known, and the angle of rotation of the cam required to achieve this motion is often also known, but where the exact movement of the follower between the beginning and end of its motion is at the discretion of the designer of the cam. As an example, a picking cam in a loom must strike the picking stick in such a way that at the end of its forward travel the picking stick is moving at a known velocity. This must occur within a predetermined angle of rotation of the cam. The exact shape of the cam, and consequently how the picking stick achieves this speed, is at the discretion of the designer of the cam. In such circumstances various cam shapes can be fitted, and one of the simplest and widely used of these is the parabolic cam.

The parabolic cam shape, as its name implies, produces a movement of the follower which follows the equation

$$x = a\theta^2 \tag{4.11}$$

where a is constant. Such a shape, while perfectly suitable for the rise portion of a follower movement, would not connect well with a falling or return section, or even with a dwell section, as can be seen from Fig. 4.11a. To join the curves smoothly together the curve follows Equation (4.11) until it reaches halfway up

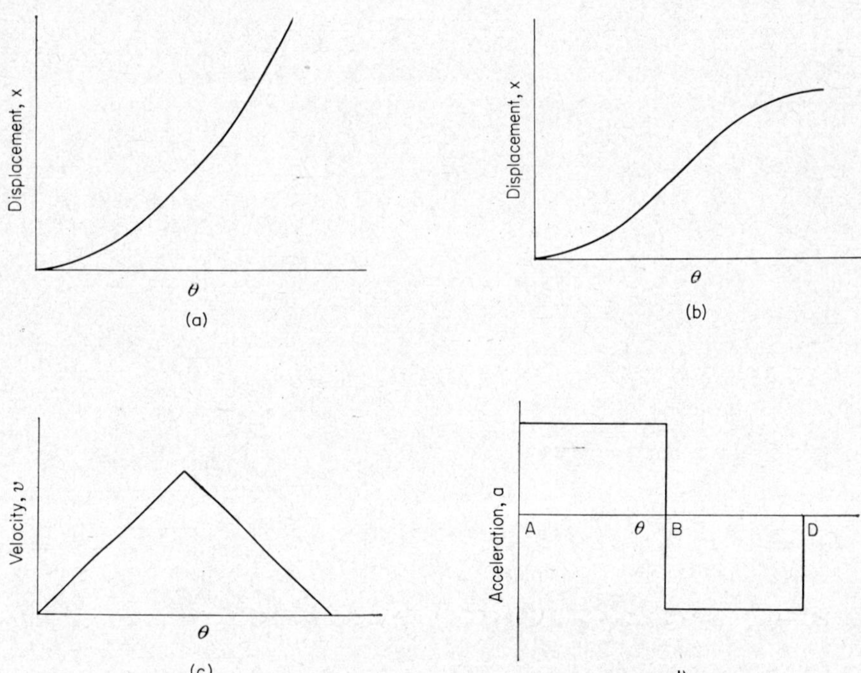

FIG. 4.11. Displacement, velocity and acceleration of the follower
with a parabolic cam.

the rise section, when the curve is replaced by its mirror image, as shown in
Fig. 4.11b. The equation for the second half of the rise portion of the curve is
given by an equation of the general form

$$(b - x) = a(\theta_0 - \theta)^2 \qquad (4.12)$$

The values of the constants a, b and θ_0 in Equations (4.11) and (4.12) are quite
simply related to the total height h that the follower must rise, and the angle ϕ
through which the cam rotates during this rise. Thus since (by symmetry)
$x = h/2$, when $\theta = \phi/2$, we can substitute in Equation (4.11), and obtain

$$\frac{h}{2} = a\left(\frac{\phi}{2}\right)^2$$

from which

$$a = \frac{2h}{\phi^2}$$

Substituting in Equation (4.12) and noting again that when $x = h/2$, $\theta = \phi/2$:

$$\left(b - \frac{h}{2}\right) = \frac{2h}{\phi^2}(\theta_0 - \tfrac{1}{2}\phi)^2 \qquad (4.13)$$

required to calculate the maximum and minimum radii of the cam disc, given that the roller is $1\frac{1}{2}$ inches in diameter.

If the cam rotates at n r.p.m. then the follower moves a distance of 10 inches for each half-revolution so that the velocity v is $2n \times 10$ (in/min), since the follower is moving linearly with time. The rotational speed of the cam, ω_1 is $2\pi n$ rad/min so that:

$$\frac{v}{\omega} = \frac{20n}{2\pi n} = \frac{10}{\pi} \text{ in}$$

Since ϕ cannot exceed $20°$, $\tan \phi$ cannot exceed $0\cdot364$, and the minimum value of c is given by:

$$0\cdot364 = \tan \phi = \frac{v}{\omega c} = \frac{10}{\pi c}$$

Hence c must exceed:

$$\frac{10}{0\cdot364\pi} = 8\cdot75 \text{ in}$$

At the start of the rise section, c will be at its minimum value, and at this point c equals the sum of the radii of the roller and the cam disc. Therefore the minimum cam disc radius is $8\cdot75 - \frac{1}{2}(1\cdot5)$ in, i.e. $8\cdot0$ in. After the rise section, c is now 10 in larger than at the start, and once again contact is made along the line of centres O_1O_2, so that the maximum cam disc radius is also 10 in larger than the minimum radius, i.e. $18\cdot0$ in.

The above analysis can also be used for another purpose. In Section 4.3 we showed how it was possible to obtain the equation for the shape of the cam disc in terms of polar coordinates if the follower was knife-edged and the required equation of motion of the cam follower was known. We shall now show how it is possible to obtain such an equation for a roller follower. Fig. 4.14 shows a line drawn on the cam disc, O_1B, making an angle θ with O_1O_2. As the cam rotates, this angle increases, and clearly as O_1O_2 is a fixed direction

$$\theta = \omega_1 t \qquad (4.25)$$

The methods given in Section 4.3 can clearly be used to obtain a relationship between c and θ, since c is the radial distance from O_1 to a fixed point on the follower. The relationship for a linear cam, for example, will be given by Equations (4.7) and (4.8), where r is replaced by c. To obtain the shape of the cam from this relationship it is now necessary to find the equation connecting b, the radial distance to the point of contact, with θ', the angle which this radius makes with the fixed line.

Clearly $\theta' = \theta + \psi$, so that if we can obtain b and ψ at any value of θ and c, the equation of the shape of the curve will be determined. This is basically simple but involves fairly lengthy calculations so that only the principle of such a calculation will be given here. Equation (4.24) gives us a value of ϕ in terms of v, ω, and c. Since the equation for the displacement of the follower is known we know that:

$$x = f(\theta)$$

and from Equations (4.9) and (4.25)

$$v = \frac{\mathrm{d}x}{\mathrm{d}t} = \omega_1 \frac{\mathrm{d}}{\mathrm{d}t}\{f(\theta)\}$$

so that

$$\tan \phi = \frac{\mathrm{d}}{\mathrm{d}t}\{f(\theta)\}/c, \qquad (4.26)$$

which can be found for all known values of θ and c. From Fig. 4.14, however, we also have

$$b \cos \psi = c - a \cos \phi \qquad (4.27)$$

and since we have already seen that

$$b \sin \psi = a \sin \phi$$

we have

$$\tan \psi = \frac{a \sin \phi}{c - a \cos \phi} \qquad (4.28)$$

The process of calculation is therefore as follows. For a given value of θ we find c from equations similar to (4.7) and (4.8). ϕ can therefore be found by substituting in Equation (4.28) which can then be used to find b from Equation (4.27). Since $\theta' = \theta + \psi$ we have found b and θ', and determined the profile of the cam.

There are simpler graphical methods for obtaining this profile but, as will be seen in the next section, such methods are not sufficiently accurate for determining the profile of cams which are to be used for high-speed machinery.

4.8 THE VIBRATION OF THE CAM FOLLOWER

In Section 4.3 we found that the linear cam was unsuitable for any application in which the cam speed was of any appreciable size, because of the effect of the very large accelerations that are produced by such a cam shape. In Section 4.4 it was shown that these large accelerations could, in some applications, be prevented by using parabolic cam shapes. If the speed of the cam is increased still further it is found that, even if the accelerations are not excessive, the parabolic cam cannot be used because it sets up vibrations in the cam follower. These vibrations are the direct result of the follower leaving the cam surface so that the follower motion is no longer dictated by the cam profile.

Fig. 4.15a shows the actual movement of a follower on a parabolic cam driven at high speed. The 'computed curve' shows the required movement of the follower. It can be seen that the vibration is so severe that the follower movement bears almost no relationship to the required movement. Fig. 4.15c shows that another cam shape produces a much better follower movement at high speeds. Why is this so?

It has been shown that the basic reason for the vibration shown in Fig. 4.15a is not the acceleration of the follower, but the rate of change of acceleration. This rate of change of acceleration, or pulse as it is called, is directly proportional to the rate of change of the inertia load. If the inertia load changes smoothly

H

then the vibration is manageable, but if the load changes suddenly then the vibration produced is large. If we look at Fig. 4.11d, which shows the acceleration of a parabolic cam, we can see that for most of the cycle the acceleration is constant, but that at three points along the cycle—at points A, B and D, where the sections join each other—the acceleration changes suddenly from a negative to a positive value, and vice versa. The pulse at these points is therefore infinite, and it is not altogether surprising that such a cam profile produces the large vibrations shown in Fig. 4.15a.

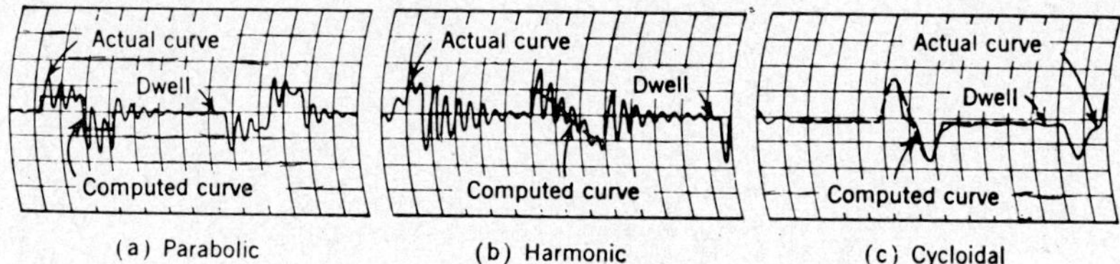

FIG. 4.15. Vibration caused by high-speed running with a. parabolic, b. harmonic, and c. cycloidal cams.
(Reproduced from D. B. Mitchell, *Mech. Eng.*, **72**, p. 467, June 1950.)

There are many possible cam profiles for producing non-infinite pulses, but one of the most important is the cycloidal cam shape. The equation for the follower movement of a cycloidal cam shape is:

$$x = m(k\theta - \tfrac{1}{2}\sin 2k\theta)$$

where m and k are two arbitrary constants. This cam shape is frequently used for high speed cams where a dwell-rise-dwell action is required. The reason for using this particular cam shape becomes apparent when we examine the velocity and acceleration of the follower produced by it. As we have already seen, the velocity and acceleration are proportional to $dx/d\theta$ and $d^2x/d\theta^2$ respectively. Hence we have:

$$v \propto \frac{dx}{d\theta} = mk(1 - \cos 2k\theta)$$

and

$$m \propto \frac{d^2x}{d\theta^2} = +2mk^2 \sin 2k\theta$$

The displacement, velocity and acceleration of this cam shape are illustrated in Fig. 4.16. It can be seen that the cycloidal cam produces a single rise to a dwell, and in so doing produces a motion with a zero velocity at the beginning and end. The acceleration changes are smooth at all points, and the acceleration is zero at the beginning and end of the motion. Thus there is no sudden change in the acceleration at any point.

The two arbitrary constants m and k are directly related to the height that the cam lifts the follower and the cam angle over which this change takes place. To show how m and k are determined let us consider a specific example. A cam is to be constructed to raise a follower a distance of $\tfrac{1}{2}$ inch in $60°$ movement of the cam. We wish to find the equation of the cycloidal cam which will produce such a movement. The general equation of the cycloidal cam is:

$$x = m(k\theta - \tfrac{1}{2}\sin 2k\theta)$$

θ is measured from the beginning of the motion of the follower, i.e. $x = 0$ when $\theta = 0$. Then since the follower is to rise through $\tfrac{1}{2}$ in while the cam rotates

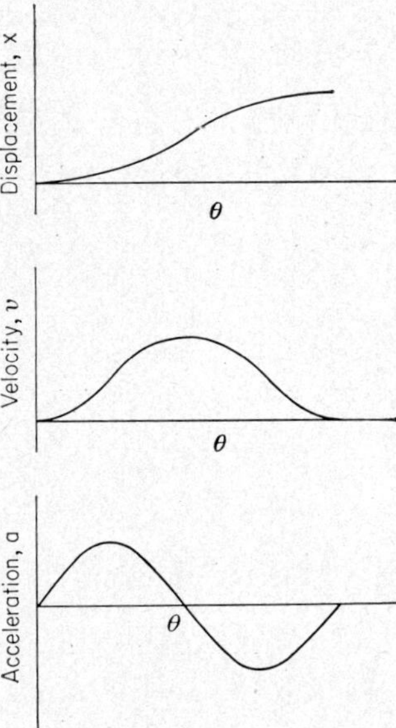

FIG. 4.16. Displacement, velocity and acceleration of the follower with a cycloidal cam.

through 60° we must have $x = \tfrac{1}{2}$ when $\theta = 60° = \pi/3$ radians. Substituting the values in the above equation:

$$\tfrac{1}{2} = m\left(\frac{k\pi}{3} - \tfrac{1}{2}\sin\frac{2k\pi}{3}\right) \tag{4.29}$$

This is one equation relating m and k. To determine these two quantities we require another equation. This is obtained from the condition that we require the acceleration to be zero at the end of the motion (the basic equation already ensures that the acceleration is zero at the start of the motion). Since the acceleration is proportional to $\sin 2k\theta$, we must therefore have

$$\sin\left(\frac{2k}{3}\right)\pi = 0$$

which implies that

$$\frac{2k}{3}\pi = \pi$$

or

$$k = \frac{3}{2}$$

neglecting the solution $k = 0$. Substituting the value of k in (4.29) to find m we obtain:

$$\frac{1}{2} = m \times \frac{3}{2} \times \frac{\pi}{3}$$

whence:

$$m = 1/\pi \text{ in}$$

The required equation is therefore

$$x = \frac{1}{2\pi}(3\theta - \sin 3\theta)$$

The cycloidal shape, as has already been mentioned, is not the only shape which produces a smooth acceleration, and Table 4.2 gives a summary of some of the most commonly used cam shapes, together with some of their most important properties. Many of these properties are self-explanatory, but column 2 gives a measure of the relative size of the cam. This value represents the cam angle through which the cam moves to accomplish a fixed movement relative to the angle through which a linear cam must be moved, the maximum pressure angle and velocity of all the cams being the same. As we have noted before, the linear cam is the most compact cam in the sense that it raises the follower through the required distance in the smallest angle of rotation, if the maximum cam slope, $\mathrm{d}x/\mathrm{d}\theta$, is fixed. Column 2 shows that all other cam shapes require much larger angles to accomplish the same task. This is a direct result of the motion which is produced by them being more gradual and less jerky than that produced by the linear cam. The values given in column 2 can be looked at in another way, for they also represent the ratio of the maximum pressure angle of the given cam to that of the linear cam—if in all cases the cam action has been compressed into the same angular movement of the cam. As was noted previously, knitting cams must have their action compressed into a small total cam length, and column 2 shows that if non-linear cams are used for this purpose very much larger pressure angles will result.

Table 4.2

Basic Curve	Cam Size Ratio	Displacement Equation $x =$	Motion		Application
			Velocity	Acceleration	
Linear	1	$m\theta$	Constant	Infinite at ends	Knitting machines
Simple harmonic	1·6	$m(1 - \cos k\theta)$	Zero at ends, maximum in the centre	Moderate at ends, zero in the middle	General purpose shape for moderate speed cams
Parabolic	2	$m\theta^2$	Uniform increase	Constant	Picking cams
Cycloidal	2	$m(k\theta - \tfrac{1}{2} \sin 2k\theta)$	Low velocities at start and finish	Smooth curve, zero at start and after rise	General purpose high speed cam shape
Double harmonic	2	$m\{(1 - \cos k\theta) - \tfrac{1}{4}(1 - \cos 2k\theta)\}$	Lowest velocities at start of all common cam shapes	Smooth curve zero at start and after return to original position	Particularly suitable for high speed dwell-rise-return-dwell cams

The behaviour of the various cam shapes differs widely but, except for the linear cam, their profiles are not very different. Table 4.3 shows the follower displacement of three cam profiles in addition to the linear cam. To make comparisons easier they have all been chosen so that they move the follower through one unit of distance in 1 radian of cam rotation.

Table 4.3

Cam Angle	Linear Cam	Parabolic Cam	Simple Harmonic Cam	Cycloidal Cam
0	0·000	0·000	0·000	0·000
0·125	0·125	0·031	0·038	0·012
0·250	0·250	0·125	0·146	0·090
0·375	0·375	0·281	0·308	0·262
0·500	0·500	0·500	0·500	0·500
0·625	0·625	0·719	0·692	0·738
0·750	0·750	0·875	0·854	0·910
0·875	0·875	0·969	0·962	0·988
1·000	1·000	1·000	1·000	1·000

As can be seen from Fig. 4.15 the small differences in shape between the parabolic and cycloidal cam have an extremely large effect on the high-speed behaviour of the cam. It follows, therefore, that high-speed cams must be very accurately ground if they are to be successful in such high-speed applications. In addition, considerable care must be taken in lubrication to minimise wear as far as possible.

Many textile machines made in the nineteenth and early twentieth centuries were relatively slow-moving and the cam shapes used were not critical. It became common practice for the overlooker and loom tuner to hand grind the cams on his machine to attempt to achieve maximum performance. Table 4.3 shows that with high-speed machinery such tampering with the machine can only be disastrous.

4.9 THE EFFECT OF LINKAGE ELASTICITY ON THE FOLLOWER MOVEMENT

We have so far considered the movement of the follower at the point where it meets the cam disc, and have assumed, implicitly, that the movement of the other end of the follower, where it operates on the machine, is the same. This will only be true, however, if the follower is made of inelastic material. We have already seen that the acceleration of the cam is never constant over the whole cycle of its movement, and consequently the inertia forces on the follower change. As the forces change, the physical dimensions of the follower alter due to the elasticity of the material. As a result, the movement of the follower at the cam disc is not the same as the movement of the machine part on which the cam and its follower are acting. This might appear at first sight to be an academic point and that the difference between the follower movement due to the cam shape, i.e. the nominal movement, and the actual movement is rather small. A simple analysis of the movement of a shuttle when driven by a linear picking cam shows that the effect of the follower elasticity is very large and, in fact, determines the cam shapes used in picking mechanisms.

Fig. 4.17 shows a picking stick and shuttle with a nominal position at time t at a distance s from its initial position (assume that when t is zero the shuttle is at its initial position). The inertia of the shuttle will exert a force on the picking stick, thereby bending the picking stick backwards so that its distance x from the initial position at time t is less than s. Since the picking stick is elastic, the force exerted on the stick will be directly proportional to the distance through

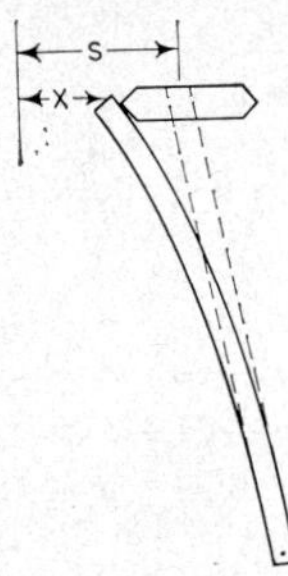

FIG. 4.17. Picking stick bent by inertia force.

which the stick has been bent, i.e. $(s - x)$. Hence the inertia force on the stick I is given by:

$$I = k(s - x)$$

where k is the spring constant of the stick, the value of which will be considered later. If we neglect the mass of the picking stick, the inertia force equals the mass of the shuttle M times its acceleration. Now the acceleration of the shuttle is the rate of change of its velocity, and the true velocity of the shuttle is dx/dt (note that the nominal speed is ds/dt). Hence

$$I = M\frac{d}{dt}\left(\frac{dx}{dt}\right) = M\frac{d^2x}{dt^2}$$

and therefore

$$M\frac{d^2x}{dt^2} = k(s - x) \tag{4.30}$$

This differential equation can be solved in terms of the variation of x with time if we know how s varies with time. If we assume that the cam is linear, then

$$s = bt$$

and substituting for s in Equation (4.30) and rearranging we obtain:

$$\left(\frac{1}{n^2}\right)\frac{d^2x}{dt^2} + x = bt \tag{4.31}$$

where $n^2 = M/k$; n has been termed the alacrity of the picking stick.

The general solution of this differential equation is

$$x = bt - A \sin nt \tag{4.32}$$

where A is an integration constant whose value will be found later.

That this is a solution of Equation (4.31) can readily be seen by finding the second derivative of x. We have

$$v = \frac{dx}{dt} = b - An \cos nt \tag{4.33}$$

and

$$a = \frac{d^2x}{dt^2} = + An^2 \sin nt \tag{4.34}$$

Dividing (4.34) by n^2 and adding to (4.32) we have:

$$\frac{1}{n^2}\frac{d^2x}{dt^2} + x = + A \sin nt + bt - A \sin nt = bt$$

which is the original equation, (4.31).

Equations (4.32), (4.33) and (4.34) would, in general, represent the oscillation of the picking stick about its nominal position. But in this case it is impossible for the shuttle to have a negative acceleration, since it is not attached to the picking stick, and therefore no possibility exists of pulling the shuttle back to produce a negative acceleration. In fact, when such a negative acceleration is required the shuttle leaves the picking stick and flies across the race. In addition, the velocity at time $t = 0$ is zero, since the movement commences at time $t = 0$.

Hence we have that $v = 0$ when $t = 0$, and therefore, from Equation (4.33)

$$b - An \cos 0 = 0$$

or

$$A = b/n$$

The acceleration a, given by Equation (4.34), first becomes negative when $\sin nt$ is negative, i.e. when $nt = \pi$. The terminal velocity v_t is therefore obtained by substituting this value for nt in Equation (4.33), to obtain

$$v_t = b - \frac{b}{n}.n \cos \pi = 2b$$

The nominal velocity of the picking stick is ds/dt which equals b. Hence we see that as a result of the elasticity of the picking stick the actual velocity is twice the nominal velocity. The effect of the elasticity of the follower is clearly very important in this case.

To show how this result comes about in more detail, Figs. 4.18a, b and c show the nominal and actual displacement, velocity and acceleration of the shuttle. It can be seen that the initial infinite nominal acceleration has been smoothed out by the elasticity. This has only been possible because the shuttle has lagged behind its nominal position. It catches up with its 'correct' position when the acceleration has returned to zero. At this point, however, the velocity is twice the nominal velocity, and if the stick were to remain in contact with the shuttle, the shuttle would have to be decelerated. This is not possible, so the shuttle flies across the race at twice its nominal speed.

It should be noted that in this rather exceptional case the linear cam does not appear to have any disadvantages. But this is only because the shuttle can leave the cam follower. If the follower were attached to some fixed piece of machinery, the oscillation predicted by Equation (4.32) would continue and could make the use of such a cam impossible, as was shown in Fig. 4.15.

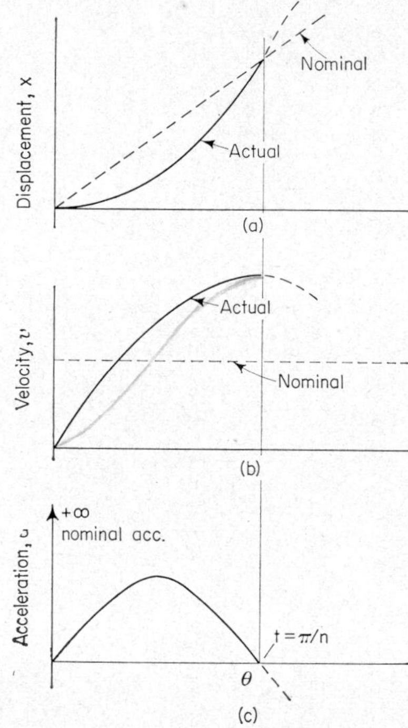

FIG. 4.18. Nominal and actual displacement velocity and acceleration of the shuttle.

In practice, it has been shown that even when used for picking, the linear cam has a major disadvantage. We have already noted that the shuttle leaves the picking stick when $nt = \pi$, or $t = \pi/n$, which, on substituting for t in Equation (4.32), shows that when this happens the shuttle has moved a distance x_m, given by

$$x_m = b\left(\frac{\pi}{n} - A \sin \pi\right) = \frac{b\pi}{n} \qquad (4.35)$$

The distance that is available for accelerating the shuttle to its ultimate speed is the size of the shuttle box. Equation (4.35) shows that since the final speed is decided by the loom design, which fixes b, the only way we can vary the distance x_m is by varying the alacrity n. For a given shuttle mass, this implies that we need to alter the elasticity of the picking stick. The more elastic the stick, the larger is x_m. It is found, however, that there is no usable material with sufficient elasticity to make x_m as large as the available distance. The same restriction does not apply to other cam shapes, which can be designed in such a way that the acceleration of the shuttle occurs over the whole length of the shuttle box. As a result, lower accelerations of the shuttle, and consequently smaller forces acting on it, are obtained by using cam profiles which do not produce a linear movement. Cam profiles have been devised enabling the acceleration to be a minimum. Table 4.4 shows the reduction in acceleration obtained by using such cam profiles in a shuttle box of standard size.

Table 4.4

Equation of Cam Movement	Maximum Acceleration (ft/s^2)
$s = bt$	2175
$s = ct^2$	1408
$s = et - f \sin gt$	1029

where b, c, e, f and g are all constants. The last cam profile was obtained by finding the most suitable values for the actual shuttle movement, and then working back to find which cam profile would give this actual movement.

This method is also used in general for eliminating vibration in cams in which the follower is firmly attached, through an elastic member, to the rest of the mechanism. It was shown in Section 4.8 that if vibrations are to be reduced when using inelastic followers, it is necessary to ensure that no sudden change occurs in the acceleration, or in other words, that the pulse is always finite. It has also been shown that if elastic followers are used it is necessary to ensure that the pulse of the actual, as well as the nominal, movement must be made finite. For details of the methods used in designing such cams the reader is referred to the paper on polydyne cams given in the bibliography at the end of this chapter.

4.10   THE USE OF POSITIVE CAMS

The cams that have been considered thus far require that contact is maintained between cam and follower by the dead weight of the follower and machine parts or, when this is impossible, by means of a spring. It is obvious that when vibrations are set up in such a system the follower can very readily leave the cam. The amplitude of the vibration can be decreased by using positive cams such as those described in Section 4.1. While the amplitude of the vibration is decreased, undesirable side effects will still result. For example, the needles in a knitting cam will, when conditions are unsuitable, bounce backwards and forwards between the two faces of the cam, producing score marks at the points where the needles strike the cam. As these marks become deeper with continuous use of the cam, they result in the jamming of the needles. It is consequently just as important to use the best possible cam profiles on positive as on negative cams.

The simple positive cam shown in Fig. 4.6 is widely used in textile machinery—for several examples of such applications see Section 5.7—but it has a disadvantage not shared by the simpler negative cam. In the negative cam the roller of the follower is always rotating in the same direction whether the follower is being raised or lowered. In the type of positive cam shown in Fig. 4.6 the roller must change its direction of rotation when the drive changes from the one side of the groove to the other. It has been found that at high speeds the roller will continue to rotate in the wrong direction for a relatively long time due to the inertia of the roller. As a result the rollers 'skid' along the cam surface producing excessive wear and eventual failure.

The alternative cam system shown in Fig. 4.7 does not have this disadvantage, as the rollers are always rotating in the same direction. Such a double cam is an expensive machine part consisting of two very accurately ground, hardened discs, to which the two rollers on the follower must be fitted very carefully to prevent jamming or excessive backlash. In addition, to maintain the accuracy of the surfaces, care must be taken in lubrication to prevent excessive wear. It can be seen, therefore, that the cam systems, which at first sight appear to be very simple ones, become expensive and critical parts of a machine, especially if it is being operated at high speeds. As will be seen in the next chapter, they can be replaced by the more robust and, on the whole, simpler linkage mechanisms. In so doing, however, we lose the versatility of the cam system which enables us to reproduce any required movement, an ability possessed by no other form of mechanism.

Suggested Further Reading

A. H. ROTHBART. *Cams—Design Dynamics and Accuracy*. John Wiley, 1956.
D. A. STODDART. 'Polydyne Cam Design', *Mach. Des.* **25**, p. 121, January 1953; p. 146, February 1953; p. 149, March 1953.

CHAPTER 5

LINKAGE MECHANISMS

5.0 INTRODUCTION

As will be seen later in this chapter, linkage mechanisms can be of many different kinds but fortunately the simplest type is also the most important. The essential features of this class of linkage mechanism are shown in Fig. 5.1. The mechanism consists essentially of four rods. Each rod has a turning or hinged connection at each end which connects it to the next rod in the chain of the linkage mechanism. One of the rods, or links, is fixed in space; one, known as the driving arm, is usually rotated at a constant speed, and the final drive is taken from some position on one of the other two rods. Fig. 5.1 is an abstraction of many mechanisms, several of which will probably be familiar to the reader.

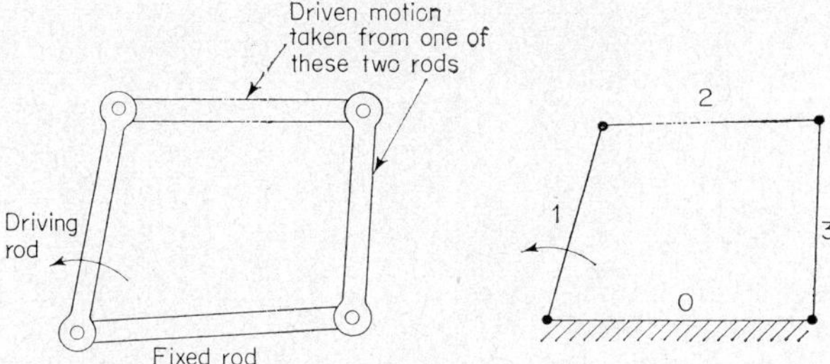

Fig. 5.1. View and diagrammatic representation of a four-bar linkage.

For example, Fig. 5.2 shows a sley being driven by a crank. In this example, the small arm of the crank AB is the constantly rotating driving link, while the second link is the connecting rod BC, from the crank to the sleigh. The sley itself, CD, is the final driven link pivoted in the frame of the loom. The loom frame, which may be represented by AD, is the fixed link, since it has the pivot or turning pair of both the sley and the crank arm fixed in it.

This four-bar linkage is only one example of a large number of classes of linkage mechanisms; it consists of turning pairs only. It is possible to use sliding pairs in a linkage mechanism, and one or two of these will be considered later in this chapter. It is, however, true to say that they do not play a large part in textile machinery, and they will not be considered in any great detail.

A further and more important extension of the four-bar linkage is the use of more than four bars in the mechanism. In considering such possible extensions, two points must be kept in mind. Firstly, it must be possible for the bars,

whence

$$\sin \gamma = \left(1 - \frac{b^2}{c^2} \cos^2 \beta\right)^{\frac{1}{2}}$$

and therefore

$$x = b \sin \beta + c\left(1 - \frac{b^2}{c^2} \cos^2 \beta\right)^{\frac{1}{2}}$$

Since b/c is less than 1, the square root can be expanded as a binomial series, giving

$$x - c = b\{\sin \beta - \tfrac{1}{2}\left(\tfrac{b}{c}\right) \cos^2 \beta + \tfrac{1}{8}\left(\tfrac{b}{c}\right)^3 \cos^4 \beta - \ldots\}$$

But $\beta = \omega t$; hence $x - c$ is given by:

$$x - c = b\{\sin \omega t - \tfrac{1}{2}\left(\tfrac{b}{c}\right)^2 \cos^2 \omega t + \ldots\} \qquad (5\cdot3)$$

The velocity and acceleration of the point C can very readily be found by differentiating this expression for x with respect to t. Thus we find:

$$v = \frac{dx}{dt} = b\omega\{\cos \omega t + \tfrac{1}{2}\left(\tfrac{b}{c}\right)2 \sin \omega t \cos \omega t - \ldots\}$$

$$= b\omega\{\cos \omega t + \tfrac{1}{2}\left(\tfrac{b}{c}\right) \sin 2\omega t - \ldots\} \qquad (5\cdot4)$$

and

$$a = \frac{dv}{dt} = - b\omega^2\{\sin \omega t + \left(\tfrac{b}{c}\right) \cos 2\omega t + \ldots\} \qquad (5\cdot5)$$

A case of considerable interest is that in which b is much smaller than c, when all the terms, except the first, in the above expressions can be neglected. It can then be seen that the resulting equations are the equations for simple harmonic motion, obtained in Section 4.2. As a first approximation it can be assumed that all four-bar linkages produce a simple harmonic motion. The properties of this motion will therefore be considered in some detail.

5.2 SIMPLE HARMONIC MOTION

The general relationship between the displacement, velocity and acceleration of a body performing simple harmonic motion is shown in Fig. 5.6a, b and c. All the movements are smooth and the pulse of such a movement is never infinite. It is therefore capable of being used at high speeds without excessive vibration. An interesting point about this motion is that the acceleration is always opposite in sign to the displacement. Thus, consider a shuttle resting on a race which is moving with simple harmonic motion. As soon as the race passes its centre position and moves towards back centre, the acceleration of the shuttle is outwards away from the loom. There must therefore be a force acting on the shuttle in this direction. This force is applied to it by the reed. In other words, the shuttle is pressed against the reed during its flight simply due to the movement of the sley. It is not necessary to provide any other means for ensuring that the reed and shuttle are in contact. It should be noted that if the shuttle were fired across the race when the sley was at front centre (this would only be possible if there was no yarn in the loom) the shuttle would fly out of the

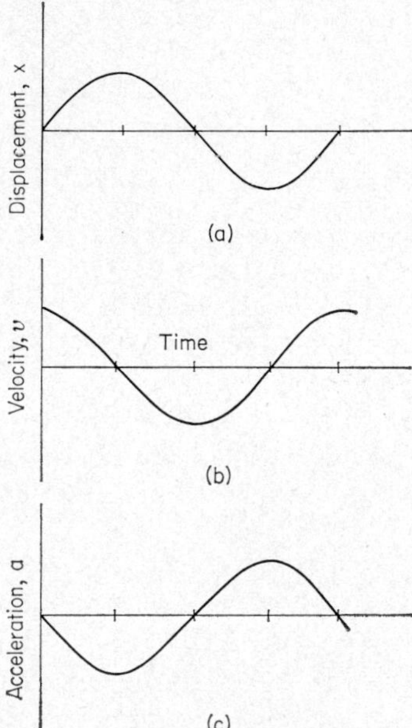

FIG. 5.6. Displacement, velocity and acceleration of the simple harmonic motion.

loom as there is then no means of keeping it in place, except for the friction between the shuttle and the race.

As a general motion for use in many mechanisms the simple harmonic motion has the advantage that the period of time spent at the two end quarters of the motion are considerably greater than that spent in the two centre quarters. To consider a practical example, the shuttle can only be propelled across the race while the race is near the back centre; otherwise the shuttle would be trapped in the warp. The fraction of the total movement of the race during

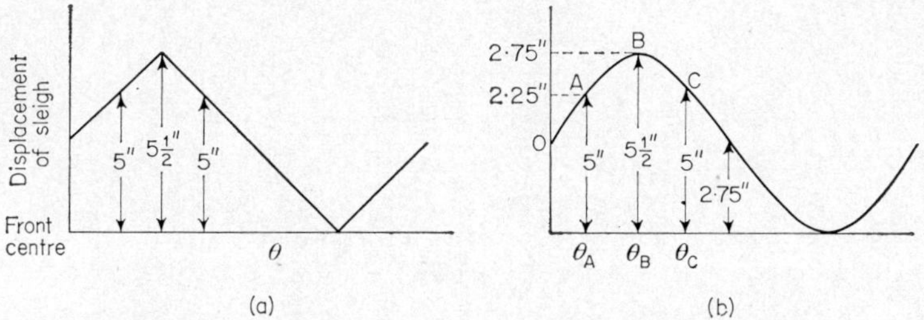

FIG. 5.7. Calculation of dwell clearance time on a loom.

I

which the shuttle can be fired is about 10 per cent, but the fraction of the total
time available is about 20 per cent.

To clarify this point let us calculate the fraction of the total cycle which can
be used for propelling a shuttle across a race (*a*) when the race is being propelled
by a linear cam, and (*b*) when it is being propelled by simple harmonic motion.
We shall suppose that the sley moves through a distance of $5\frac{1}{2}$ inches and that the
shuttle can be propelled when the race is more than 5 inches from the front centre.

Fig. 5.7 shows the movement of the sleigh both for a linear cam and for a
simple harmonic motion. It can be seen that the fraction of the total time
available on a linear system is $(5\frac{1}{2} - 5)/5\frac{1}{2}$ times the whole cycle time. Express-
ing this as a fraction of the total angular rotation of the crankshaft we obtain

$$\frac{0\cdot5}{5\cdot5} \times 360° = 32\cdot7°$$

For the simple harmonic motion (Fig. 5.7b) we first note that the motion represented by the equation

$$x - c = b \sin \theta$$

gives values of $x - c$ which range from $+b$ to $-b$. Hence, since the movement
of the sley is $5\frac{1}{2}$ in, we must have $2b = 5\cdot5$, or $b = 2\cdot75$ in.

The shuttle can be fired when it is more than 5 inches from front centre, i.e.
between the points A and C in Fig. 5.7b. The sley passes through these
points at times represented by the angular displacements θ_a and θ_c. Hence the
time available is proportional to $\theta_c - \theta_a$, and this is the quantity we must
calculate.

Now the value of x at A is given by

$$x - c = 2\cdot25$$

Substituting this value in the above equation for simple harmonic motion,
gives

$$2\cdot25 = 2\cdot75 \sin \theta_a$$

or

$$\sin \theta_a = 0\cdot8183$$

which gives

$$\theta_a = 54\cdot9°$$

By symmetry

$$\theta_c = 180° - 54\cdot9° = 125\cdot1°$$

Therefore

$$\theta_c - \theta_a = 125\cdot1° - 54\cdot9° = 70\cdot2°$$

This is the total angular movement available for firing the shuttle, and, as can
be seen, it is more than twice as great as that for the linear cam.

A linear cam would not be used for such a purpose, however, since it is
possible to design a cam which will produce a large back-centre dwell on the
sley. Theoretically, the dwell could be nearly 360° but to prevent excessive
speeds and accelerations of the reed and sley, practical cam systems for driving

the sley usually have a back-centre dwell of about 150°. This is much larger than that available if a linkage system is used.

This simple example illustrates one of the basic differences between a cam system and a linkage system. We have already seen that both cam and linkage systems can be used to obtain a simple harmonic motion, but while the cam system can be designed to obtain *any* required motion, this is not true of a linkage system. It will be shown in the subsequent sections of this chapter that motions differing considerably from a simple harmonic motion can be obtained with a linkage system. However, these can only approximate to the required motion, and the closer the approximation required the greater is the complexity of the linkage system. In addition, a linkage system can never be constructed to have a true dwell. The motion can be very small over a considerable period, but it will never cease completely. This does not usually matter greatly as a complete stoppage is not usually necessary, but the fact that very large dwells are possible in cam systems provide them with a major advantage over linkage systems.

In view of these major disadvantages of the linkage system, why are they so widespread and why are they displacing cam systems in many modern high-speed machines? Their use arises directly from the fact that they are basically simpler and cheaper to manufacture, more robust in operation, and capable of operating at very high speeds without excessive vibration. The main rubbing surfaces in a linkage mechanism are bearings, where contact is made over a large area, whereas the contact in a cam mechanism is along a line. The linkage mechanism owes its robustness to this essential fact. In addition, the acceleration changes in a linkage mechanism are, by the very nature of the mechanism, smooth and without any sudden jerks. There is therefore no necessity to produce the accurately machined surfaces needed on cam mechanisms. Furthermore, it is possible to use linkage mechanisms in places where only a minimum amount of lubrication is available, by the use of ball or roller bearings in the pivots. As a result of all these advantages, considerable efforts have been made to find means for modifying the simple harmonic motion to obtain some more general motions. The simplest of these motions are obtained by the use of short connecting links. The effect of such shortened links will be considered in the next section.

5.3 MODIFIED SIMPLE HARMONIC MOTION

The simple harmonic motion is obtained by making the rotating link shorter than the connecting link, these two links being much shorter than the remainder. The simplest modification to the simple harmonic motion is obtained by making the connecting link about the same size as the rotating link (note that it cannot be equal to, or shorter than, the rotating link). To show the effect of such a modification let us consider a linkage in which the connecting link is 1·2 times as large as the rotating link. The movement of the rocking link is then given by Equation (5.3) where $b/c = 1/1·2$. Fig. 5.8 shows the displacement versus the angle θ for a simple harmonic motion and for the linkage with a ratio of $b/c = 1/1·2$. It can be seen that the use of a shorter connecting rod has resulted in a sharper curve towards front centre and a flatter curve at back centre. The dwell at back centre is thereby increased. As an illustration of the change we can consider the example calculated in Section 5.2 where it was shown that the shuttle could be propelled across the race while the sley moved through 70° if the sley movement was simple harmonic. If we repeat this calculation for the linkage system with the short cross-link, the motion of which is shown in Fig. 5.8, it is found that this angle is increased to 115°. This increase in dwell time is very useful, and it is common practice to use short connecting links in the sley movement to take advantage of this fact.

Further modifications of the available movement can be made by using links which are all of similar lengths, and by taking the final movement from some

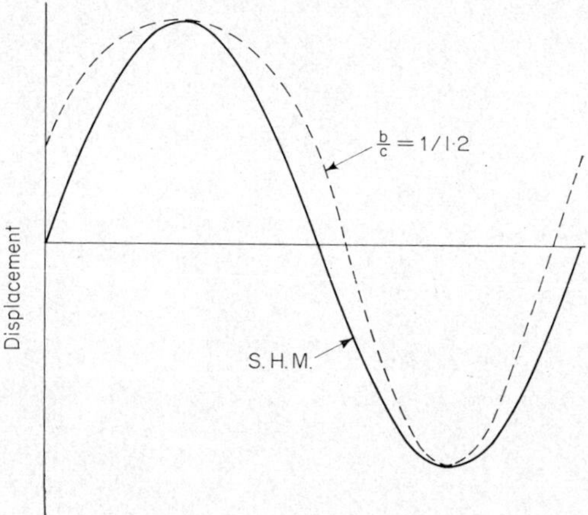

FIG. 5.8. Effect of crank-arm ratio on dwell.

point other than on the rocking arm. The analysis of the general four-bar linkage can be obtained by solving Equations (5.1). This is a lengthy and laborious process but can be carried out numerically without much difficulty. Thus, for example, at any value of β the values of γ and δ can be found from Equations (5.1) by solving two simultaneous quadratic equations. Knowing δ, the path traced out by a point on the rocking link can readily be found.

If the velocity and acceleration of this point are needed they can also be found from Equations (5.1) by differentiating with respect to time. Thus, differentiating, and noting that $d\beta/dt = \omega$, we have:

$$\omega b \cos \beta + c \cos \gamma \frac{d\gamma}{dt} = d \cos \delta \frac{d\delta}{dt}$$

and

$$\omega b \sin \beta + c \sin \gamma \frac{d\gamma}{dt} + d \sin \delta \frac{d\delta}{dt} = 0$$

Multiplying the first equation by sin γ, and the second by cos γ, and subtracting we have:

$$\omega b \sin \gamma \cos \beta - \omega b \sin \beta \cos \gamma = d(\cos \delta \sin \gamma + \cos \gamma \sin \delta)\frac{d\delta}{dt}$$

or

$$\frac{d\delta}{dt} = \omega \, \frac{b \sin (\gamma - \beta)}{d \sin (\gamma + \delta)}$$

As we have already seen, the values of γ and δ can be found for any value of β, and hence the rotational speed of the rocking link is known. The acceleration of this arm can be found in a similar way, as a function of the angle β. Such calculations are simple but tedious, and they have been carried out for a large number of possible variations in the linkage size by Hrönes and Nelson, with the aid of a computer. By consulting their book, details of which are given at the end of the chapter, the behaviour of any four-bar linkage can readily be found.

The movement of any point on the rocking link is always a movement along the arc of a circle, as the arm is rocking about a fixed point. A point on the connecting arm, however, will in general produce a more complex type of two-dimensional path. It is often necessary to have such a two-dimensional path. For example, the path of a bearded needle in a flat-bed knitting machine with a fixed presser is shown in Fig. 5.9. It is possible that such a movement could be

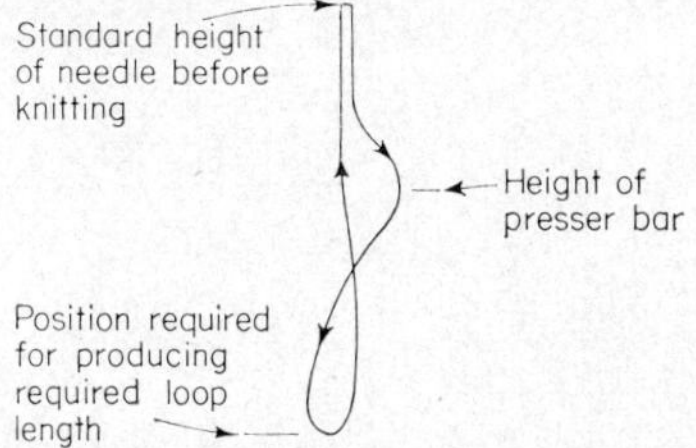

FIG. 5.9. Path of bearded knitting needle in flat bed knitting machine.

obtained by fixing the needle bar on some point on the connecting link of a four-bar linkage. This point does not necessarily have to be on the axis of the link, and many widely varying paths can be obtained in this way. Hrönes and Nelson have sketched the paths of ten specially selected points on the connecting link, which greatly assists the designer in choosing a suitable linkage for his purpose. Fig. 5.10 shows a typical example from Hrönes and Nelson's book. The original position of the linkage and the selected points are shown, together with the paths traced out by these points when the driving link is rotated. The paths are shown by means of short lines of varying lengths. These lengths have been chosen so that the length of the line represents the movement of the point during $5°$ rotation of the driving arm. By this means it is possible to find the position of the point at any time.

We have already noted in Section 5.1 that one of the links of a four-bar

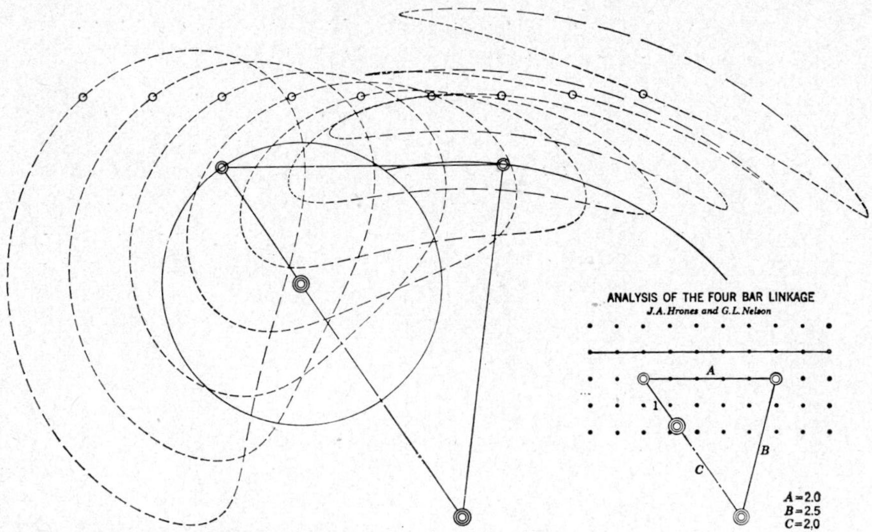

FIG. 5.10. Paths of various points on a four-bar linkage.
(Reproduced from Hrönes and Nelson, *Analysis of Four-Bar Mechanism,*
Wiley, 1951.)

linkage can be replaced by a straight slide, when the link in question is relatively long. It is possible to replace a short rocking link by a curved slide, and this is particularly useful when the linkage is being used in such a way that the required

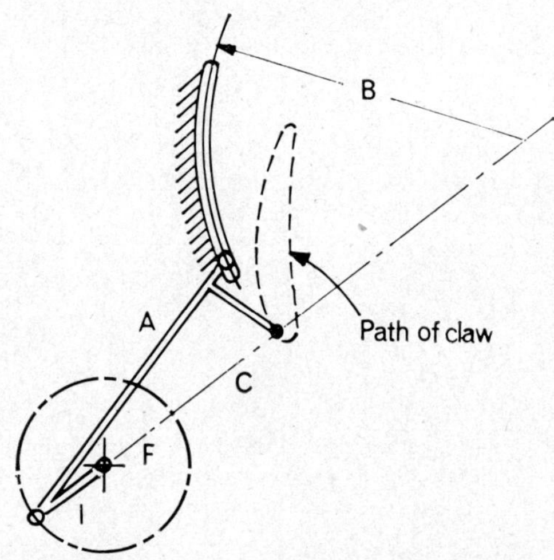

FIG. 5.11. Transfer mechanism. Since the fourth link of radius *B* would have been in the way, it was replaced by a curved slot. The path of the claw is shown by the dashed line. This motion was found with the aid of Hrones and Nelson's book [8-14, p. 70]. (Reproduced from J. S. Beggs, *Mechanism,* McGraw-Hill, 1955.)

movement is taken from the connecting link, since the rocking link will often obstruct the required path. Such a typical modification of a four-bar linkage is shown by Fig. 5.11.

Despite the wide variation in possible movements which can be obtained by careful selection, there is still a fairly large number of movements required in practice which are not covered by the four-bar linkage. A particular example of such a movement is that of the needle bar on a warp knitting machine. On the bearded needle machine the needle bar has to be moved through a complicated path which can be represented by the graph of distance moved against the angle of rotation θ shown in Fig. 5.12.

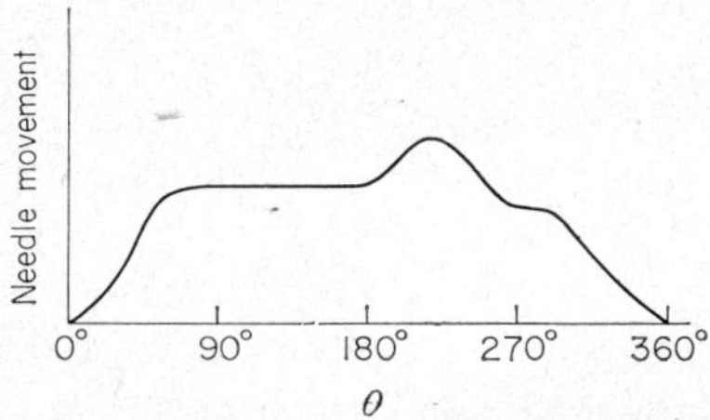

Fig. 5.12. Required needle movement on warp knitting machine.

As was mentioned in the previous chapter, this path was originally obtained by the use of positive cams. Such a drive limited the maximum speed of the needle bar and prevented the use of speeds greater than 500 r.p.m. on warp knitting machines. If the cams could be replaced by a linkage system higher speeds could be used. The required motion could not, however, be obtained by means of a four-bar linkage. For such purposes linkages containing more than four bars must be used. Their analysis is inevitably complex, and the following section merely outlines some of the multiple-bar linkages which have found important applications in textile machinery.

5.4 MULTIPLE-BAR LINKAGE

As was shown in the introduction to this chapter a multiple-bar linkage must always have an even number of links; the simplest multiple-bar linkage thus has six links. One of the simplest of these six-bar linkages is that used to obtain a double beat-up on a carpet loom. The purpose of this linkage is to move the sley backwards and forwards twice for one rotation of the crankshaft. The principle of the linkage system used is fairly simple if one remembers that in the first class of four-bar linkage systems the driving arm rotates, and the rocking arm moves backwards and forwards once for each revolution of the driving arm. If, therefore, the driving arm were to be rocked instead of rotated then by suitably choosing the size of the linkages it is possible to produce a to-and-fro movement when the driving arm moves in one direction, repeating the movement when the driving arm moves back. The rocking of the driving arm can be achieved by making it, also, the rocking arm of a normal four-bar linkage.

Fig. 5.13 shows the six-bar linkage used to obtain a double beat-up on a loom. It may be thought of as consisting of two four-bar linkages, which have two bars in common. The first of these four-bar linkages consists of the links numbered 1, 2 and 3 in the diagram, together with the frame of the machine which forms the fixed link. Link 1 is the driving link, being rotated continuously from the crank shaft. If the lengths of the links are chosen correctly the rocking link 3 will rock backwards and forwards for each revolution of link 1.

The second four-bar linkage uses link 3 as a driving link, and consists of links numbered 3, 4 and 5, the frame of the loom again being the fixed link. The rocking link of this four-bar linkage is link 5, which is also the sley. Therefore, if the lengths of the links are again chosen correctly, link 5 will rock backwards and forwards twice for each revolution of link 1, thus producing the required double beat-up.

A similar linkage mechanism, in the sense that its motion can be considered as the summation of two separate movements, is the main drive system used on

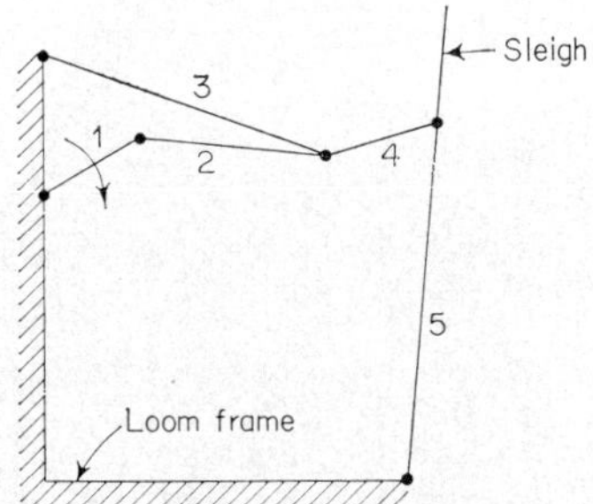

FIG. 5.13. Double beat-up mechanism.

the FNF warp knitting machine, shown in Fig. 5.14a. The main drive comes from two eccentrics which drive the connecting rods 1 and 2. These two eccentrics are driven at different speeds, the right-hand eccentric being driven at twice the speed of the left-hand eccentric. The movement of these links is conveyed to link 3, to which is attached the final link 4. This final link is pivoted at a point which is not necessarily half-way between the pivots where links 1 and 2 join link 3. Link 4 is constrained to move in a straight line as it is contained in a sleeve; this link is attached to the various members of the machine which must be operated, e.g. the guide bars, knitting bars, etc. Each of these machine members requires different movements, and these are obtained by using different variants of the linkage mechanism shown in Fig. 5.14a in which the relative position of the three pivots on link 3 is varied.

Such a mechanism can be represented by the linkage system shown in Fig. 5.14b. The eccentrics are represented by two short rotating links—the length of these links being equal to the distance between the centre of the eccentric and the centre of the shaft. As was shown previously, when discussing four-bar linkages, the sleeve constraining link 4 can be replaced by a very long link which constrains the pivot G and link 4 to move in the right direction. Note that link 4 consists of GK rigidly connected to the long arm KH.

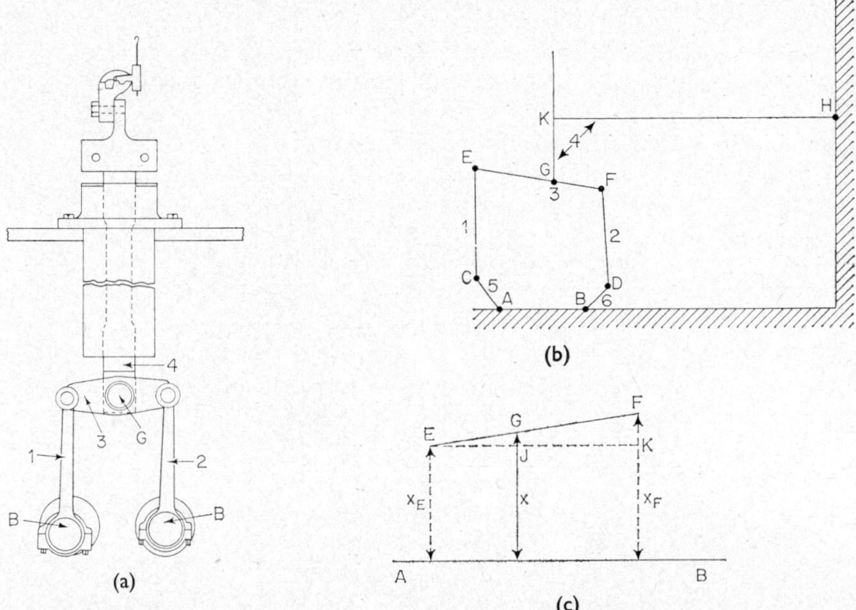

FIG. 5.14. FNF linkage mechanism for driving the needle bar.

It might be noted that there are seven links shown in Fig. 5.14b—six moving links and the fixed link. This would appear to contradict the statement previously made that a simply driven linkage must have an even number of links. In this example, however, we have two driven links. If we apply the relationship (5.1)—noting that there are eight turning points on Fig. 5.14b—we find that the number of degrees of freedom X is:

$$X = 3(L - 1) - 2J = 3(7 - 1) - 2 \times 8 = 2$$

These two degrees of freedom are required by the mechanism, as two of the links are driven. When two links are driven the number of links will always be an odd number. The behaviour of such a linkage can be analysed relatively simply. Consider first the motion of the point E. Since EG is much larger than AC, and G moves in a straight line, E must also move in a straight line, and the linkage ACE can be regarded as being similar to the linkage shown in Fig. 5.5b, in which point E is constrained by a slide. Since AC is also much smaller than EC it has already been shown that E will move with simple harmonic motion in a direction perpendicular to the fixed link AB. If we denote the distance of E from AB by x_e, then this motion may be represented by the equation

$$x_e - c_1 = a \sin \omega t$$

where a is the length of AC, c_1 is the length of EC, and ω is the rotational speed of link AC. In writing this equation we have assumed that at time $t = 0$, the link AC lies along AB.

We can find the motion of the point F in the same way, since the two links EC and DF are similar in their movements. Therefore, denoting the distance of F from AB by x_f, we have:

$$x_f - c_2 = a \sin (2\omega t + \phi)$$

where a is the length of BD, 2ω is its angular speed, and c_2 is the length of DF. The angle ϕ is introduced because at $t = 0$, the link BD does not necessarily lie along AB but, in general, make an angle ϕ with it.

We now wish to find the motion of the point G. Since KG can move only in a direction perpendicular to AB the point G can only move in this direction. Denoting its distance from AB by x, and letting $EG = b$, $GF = d$, then from Fig. 5.14c, because triangles EGJ and EFK are similar, we have

$$\frac{GJ}{EG} = \frac{FK}{EF}$$

or

$$\frac{x - x_e}{b} = \frac{x_f - x_e}{b + d}$$

which gives

$$x = x_e + \frac{b}{b + d}(x_f - x_e)$$

$$= x_e\left(\frac{d}{b + d}\right) + x_f\left(\frac{b}{b + d}\right)$$

Substituting for x_e and x_f, this gives

$$x - \frac{dc_1 + bc_2}{b + d} = \frac{ad}{b + d}\left\{\sin \omega t + \frac{b}{d} \sin (2\omega t + \phi)\right\}$$

This is the general motion produced by the FNF mechanism, and some of the possible movements that can be obtained with this mechanism are shown

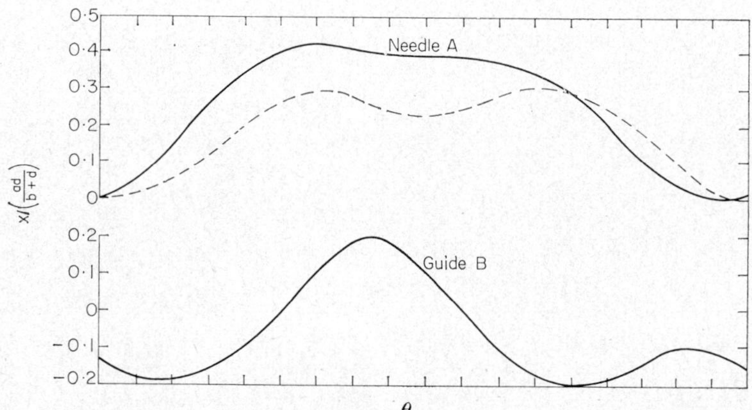

FIG. 5.15. Movement obtained by varying the parameters of the FNF linkage mechanism.

graphically in Fig. 5.15, in which $x/(ad/b + d)$ is plotted against θ, where $\theta = \omega t$. The graph A is for $b/d = 1:3$, and $\phi = 10°$, and is the movement chosen for the needle bars. Graph B is for $b/d = 1:2$, and $\phi = 180°$, and is the movement chosen for the guide bars. In one form of the machine b/d was varied by altering the throw of the cranks, and the leverage ratio was always kept $1:1$; the basic equation of its action is however the same. The movement shown in graph A shows a very much longer dwell at the highest value of x than can possibly be obtained with any four-bar linkage. This dwell is necessary to allow the guide bars to complete their movement before the knitting needles start to move down.

The bearded needle action requires, in addition to this long initial dwell, a second short dwell during the descent of the needle, to enable the presser bar to act. The FNF action cannot produce this second dwell and is therefore only applied to compound needle machines which do not use a presser. For the

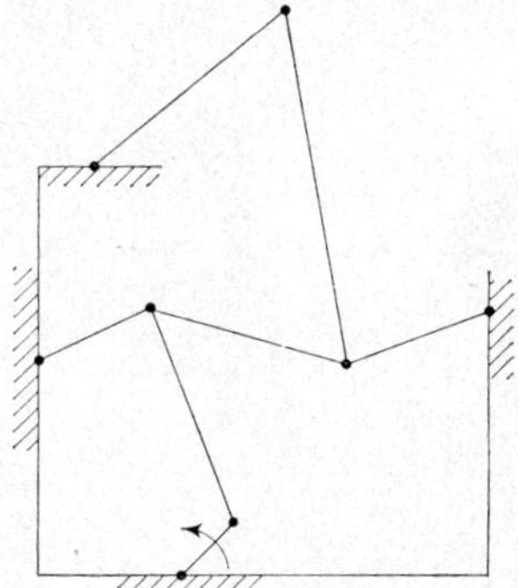

FIG. 5.16. Eight-bar linkage for needle
bar movement.

bearded needle machine a more complex linkage system is required, and one of the simplest of these is shown in Fig. 5.16.

This system employs only a single eccentric and therefore has only one driving link. It is extremely complex and cannot be analysed in the simple fashion in which we chose to analyse the previous two examples of multiple-bar linkages. There is, however, one general point that can be made in an introductory survey of this kind, which provides us with some guidance on the reason why such a complex mechanism is required.

If we are given a required movement, of a knitting needle bar for example, the cam required to produce this movement can always be found. This is, in general, impossible with a linkage system. But if we were to choose four points on the required movement (say the four points shown on graph A of Fig. 5.17) then we could find a four-bar linkage system that will produce a movement which passes through these positions at exactly the time or angular rotation specified. The movement between these chosen points is not determined. If we wish to improve the accuracy of movement by choosing six points, we will require a six-bar linkage to reproduce this movement. As the movement becomes more complicated more points will be necessary to approximate the required movement, and hence more links will be required to reproduce it. Many points are obviously needed to approximate the movement shown in Fig. 5.12, and it is therefore necessary to use many links in a mechanism which is to reproduce

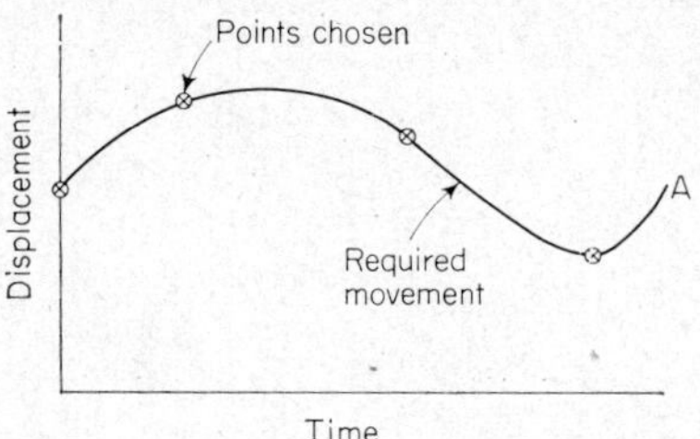

FIG. 5.17. Choosing critical points on a required movement.

such a movement. The patent literature gives examples of linkage mechanisms having as many as 72 links which have been proposed for driving needle bars on warp knitting machines.

5.5 COMBINED RATCHET AND LINKAGE MECHANISMS

A ratchet, a typical example of which is shown in Fig. 5.18, consists of a ratchet wheel and a pawl. When the pawl travels forwards (from right to left in the diagram), it engages with the flank of a tooth on the ratchet wheel, and pushes it forwards; when the pawl moves back, however, it does not engage with the teeth, and does not move the ratchet. A ratchet can therefore be used to convert a to-and-fro movement into an intermittent rotation.

To accomplish such a conversion it should be noted that the total movement of the pawl must be greater than the distance between neighbouring teeth, i.e. it must be greater than the circular pitch of the ratchet wheel. If this were not

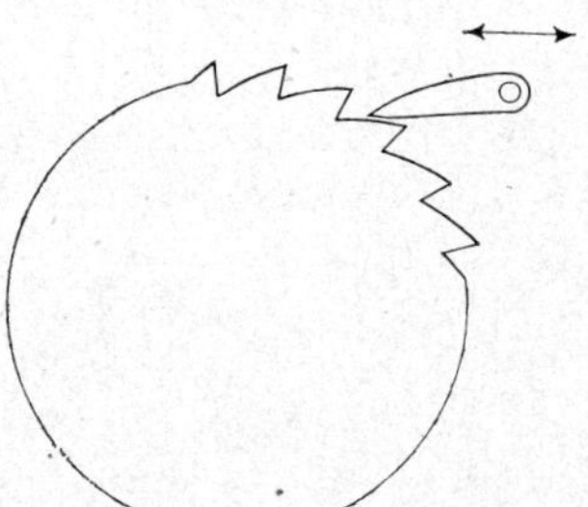

FIG. 5.18. Ratchet and pawl.

the case, the pawl, during its backward movement, would not reach as far as the next tooth, and would not therefore engage with it. As a result the pawl would merely move backwards and forwards along the curved upper face of the tooth without moving the ratchet wheel. If the pawl movement is greater than one circular pitch, but less than two circular pitches, it will move backwards sufficiently for it to engage the second tooth but not the third, and during the subsequent forward movement the ratchet wheel will be moved forwards by one pitch distance. If the movement is greater than two circular pitches but less than three, the forward movement of the ratchet wheel is two pitch distances, and so on.

This principle is often used in take-up drives, which must be rotated very slowly but need not be rotated continuously. The pawl is moved backwards and forwards by means of a simple linkage system driven directly from the main shaft of the machine. If this movement is adjusted so that the pawl movement lies between one and two circular pitches of the ratchet teeth, then for each rotation of the main drive shaft the take-up mechanism will be moved forward one tooth. Thus if there are n teeth on the ratchet wheel it will take n revolutions of the main shaft to rotate the take-up roller through one revolution. This speed reduction can be reduced still further by using reducing gear trains, either between the main shaft and the pawl mechanism (a procedure which is rarely

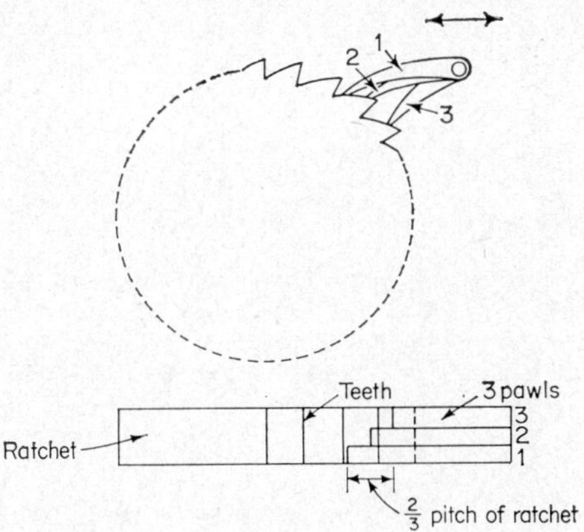

FIG. 5.19. Vernier ratchet and pawl.

used for reasons that will be apparent later), or between the ratchet wheel and the take-up roller. Another commonly used method for increasing the reduction ratio is to use a vernier ratchet.

Fig. 5.19 shows a vernier ratchet with three pawls which are moved backwards and forwards together. For the moment, let us assume that the distance moved lies between one-third and two-thirds of the pitch of the ratchet teeth. The ends of the pawls are displaced relative to each other along the path moved, by a distance equal to a third of the ratchet tooth pitch. The system operates in the following way. If the three pawls are moved forwards and pawl 1 is engaged, pawls 1 and 2 will remain on the flank of the tooth when the pawls are moved back again, but pawl 3 will drop down onto the next tooth. When the pawls are moved forwards again, pawl 3 will engage with the next tooth and move it forwards through one-third of a circular pitch. At the next cycle, pawl 2 will be engaged, and at the third cycle pawl 1 will again be engaged. It therefore takes three rotations of the pawl mechanism to move the ratchet wheel through one tooth space, and as a result the speed reduction becomes three times as great.

These ratchet drives have an important property which makes them particularly suitable for use in take-up drives. They can very readily be adapted to produce a varying take-up, the amount of take-up being controlled by the cloth tension in knitting machines, and by the warp tension in looms. (On looms the mechanism is being used as a let-off drive.) The basis of all such adaptations of the ratchet drive is the use of a shield which lies against the ratchet teeth and behind the pawl (see Fig. 5.20). The position of this shield is dictated by the tension; if the tension is too high (take-up mechanisms) or too low (let-off mechanisms), the shield moves forwards and covers all the teeth behind the pawl. As a result, when the pawl moves back it rides up onto the shield, and does not engage the next tooth. The ratchet wheel consequently remains stationary until the tension has changed sufficiently to move the shield past the next ratchet tooth. This simple method of operation is essentially an on-off control, since the ratchet wheel is either moving or stationary, depending on whether or not the tension has the correct value.

This principle can be modified to produce a much more gradual form of control. Consider a simple pawl ratchet system in which the backwards and

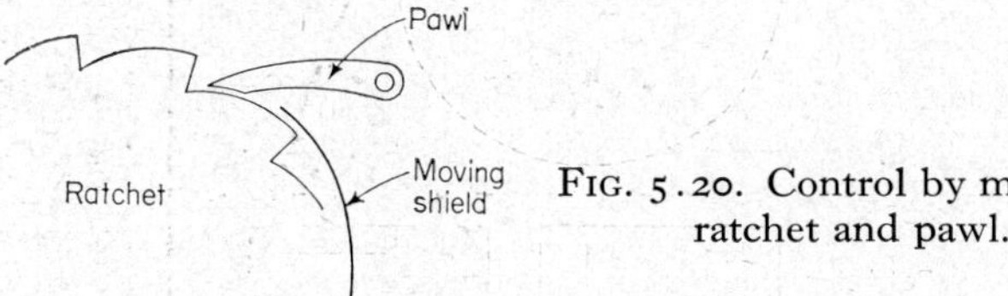

Fig. 5.20. Control by means of ratchet and pawl.

forwards movement of the pawl is between three and four ratchet pitches. This pawl will move the ratchet wheel forwards a distance equal to three teeth during each cycle. If a shield is placed behind the pawl, and this covers the third tooth behind the pawl, then the ratchet will move forward two teeth spaces. Similarly, if the shield covers the second and third teeth, the ratchet moves forwards one tooth space. The take-up or let-off can in this way be varied among 0, 1, 2 or 3 tooth spaces, depending on the tension. This will produce a much smoother variation in take-up tension. If the simple pawl is replaced by a three-pawl vernier type ratchet the required movement of the pawls and shield is decreased by a factor of 3. The pawls need to move backwards and forwards a distance between 1 and $1\frac{1}{3}$ of the tooth spacing, and the effect of the tension will be to vary the take-up or let-off among 0, $\frac{1}{3}$, $\frac{2}{3}$ or 1 teeth spaces per revolution of the crankshaft.

5.6 EXAMPLES OF COMPLEX COMBINED MECHANISMS

In this section two complete mechanisms will be examined. They consist of a combination of all the simpler basic mechanisms that have been discussed up to now. The purpose in considering these two examples is to show how even the

most complex mechanisms can be broken down into the simpler units which we
have already discussed in detail.

Fig. 5.21 shows the basic parts of the drive to the slurcock of a flat-bed
knitting machine. The slurcock must be moved across the knitting bed at a
constant speed to operate the knitting needles, and then allowed to dwell at the

FIG. 5.21. Cam and screw mechanism of flat-bed knitting machine.

far end of the bed before returning once more at a constant speed. This motion
can clearly be obtained most easily by means of a positive cam. If, however, a
cam were used to drive the slurcock directly it would have to be an extremely
large cam—about 3 ft in diameter—and what is more, it would stick out of the
side of the machine at a very awkward angle. Fig. 5.21 shows the much neater
solution used in practice. The slurcock itself is driven by means of a screw. It is
necessary therefore to rotate the screw at a constant speed, halt it for a fixed

time, and then reverse the drive. This motion is achieved by means of the mechanism shown in Fig. 5.21. The screw is attached to the gear 1, which is rotated by means of the rack 2. When the rack is moved at a constant speed the screw rotates at a constant speed. Halting the rack produces a stoppage of the screw, and so on. The rack is therefore rocked backwards and forwards by means of the positive disc cam 3. The use of rack, gear and screw has enabled the designer to magnify the movement, and turn it through a right-angle to produce a much neater mechanism. However, the essential part of this mechanism is the cam.

Fig. 5.22 shows the movement which is required from the take-off rollers of a comb. The rollers have to feed back a fibre tuft so that it is laid over the next tuft, then they must move forwards taking the new tuft with it. After one complete cycle the rollers must have moved forwards by an amount equal to the rate

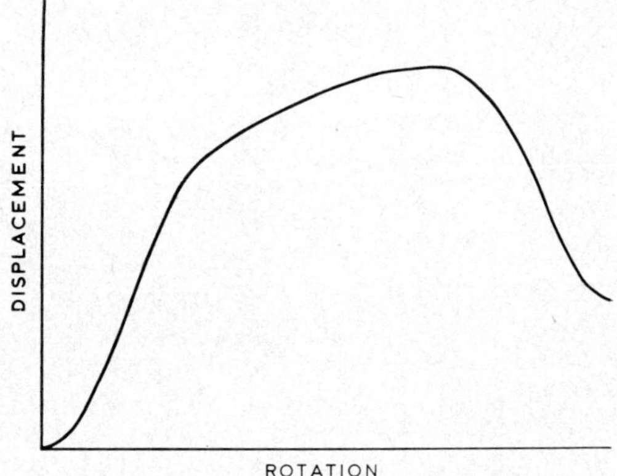

FIG. 5.22. Required movement of the detaching rollers.

of production per cycle. It can be seen, therefore, that the movement can be considered to consist of two parts; a continuous forward rotation to obtain the feed forward of the output sliver, and a forward and backward rotation superimposed upon this forward feed to obtain the correct tufting action. This addition of motion is usually obtained through an epicyclic or eccentric gear. Fig. 5.23 shows the mechanism used in the Platt's cotton comber. Gear A is eccentric with gear B and meshes with the internal teeth on the latter. The arm to which gear A is attached is rotated continuously by the main drive shaft E. Gear A is kept fixed in space by plate C when the cam follower D is not in action. Under these conditions the mechanism consists essentially of the epicyclic gear train shown in Fig. 5.23, and which comprises an internal gear, a rotating arm and a fixed gear on the arm. Such a gear train will produce a continuous rotation of the final gear B when the arm driven by shaft E is rotated. This produces the required forward movement of the detaching rollers.

To obtain the to-and-fro movement a disc cam (not shown on the figure) is used which moves the cam follower D backwards and forwards. To understand the effect of this movement, let us consider that the drive shaft E is stationary. The movement of the follower D is conveyed via the linkage mechanism to the plate C so that, as D moves up and down, the plate C rocks about the centre of the gear A. Gear A therefore rocks backwards and forwards. As a result gear B rocks backwards and forwards. When the shaft E is also rotated the two movements are added so that gear B has the required rotational motion. Teeth on the outside of gear B mesh with gear F, which is attached to the detaching rollers, so obtaining the required movement of these rollers.

These mechanisms have shown how cams, linkage and gear drives can be combined to give a desired movement. When the required movement does not vary with the type of product being manufactured, these mechanical methods of obtaining it are almost invariably the simplest and cheapest available. When the

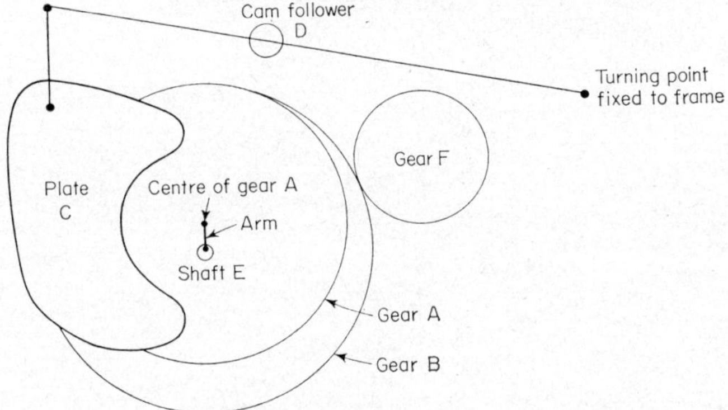

Fig. 5.23. Mechanism for driving detaching rollers.

movement needs to be varied for the different products which are made on the machine, it is often possible to consider either electrical or hydraulic means of obtaining the movements. Both rotational and translational movements can be obtained directly by hydraulic or electrical methods, and though these are, as a rule, more expensive than mechanical means, they are much more flexible. As an example, the displacement and speed of a hydraulic ram can be controlled by means of a single valve. The positioning of this valve can in turn be controlled by a small cam, so that the ram position varies in some controlled fashion. To alter the ram movement it is only necessary to replace the cam which, because it does not have to carry a large load, can be comparatively simple. An example of such a system occurs in doubling twister builder motions.

A parallel wound package can be produced from a builder motion consisting essentially of a heart-shaped or linear cam, since the required movement of the ring rail is a linear up-and-down movement. To obtain a cop-wound package requires a very gradual upward movement superimposed on a more rapid up-and-down movement. Several patented mechanical means of obtaining such a

K

movement are available, which also have facilities for producing different shaped 'bottoms' to the cop. On twisting machines, however, it may be necessary to produce parallel wound packages on some occasions, and different types of cop-wound packages on others. A hydraulic mechanism is the only feasible method of obtaining such a versatility, and is sometimes used for this purpose. Electrical and hydraulic mechanisms are still rare in the textile industry, but it is as well to remember that all the mechanisms described in this book can be replaced by non-mechanical mechanisms, and it is very probable that we shall see many more of them in the future.

Suggested Further Reading

J. A. Hrönes and G. L. Nelson. *Analysis of the Four-Bar Mechanism*. Wiley, 1951.
A. S. Hall. *Kinematics and Linkage Design*. Prentice Hall, 1961.

CHAPTER 6

SELECTION MECHANISMS

6.0 INTRODUCTION

One of the unique features of textile machinery is the extensive use made of selection mechanisms. These are almost invariably mechanical in action, and are frequently very complex. As a result their study forms a considerable part of the usual textile technology course. There would be little purpose, therefore, in dealing with these mechanisms in detail, especially as the wide range of machines which employ them would make an exhaustive, detailed study impracticable. It is proposed, therefore, to deal primarily with the principles of selection mechanisms, using examples from the various fields in which such mechanisms are used, to illustrate the principles. Also, where possible, we shall compare the different methods used in different sections of the industry.

The need for selection mechanisms arises largely from the desire to produce patterned fabrics of one form or another, and as a result such mechanisms are mainly to be found on weaving or knitting machines. The mechanism is required to select the weaving or knitting elements according to some predetermined plan so as to produce the desired effect. To take a specific example let us consider the weaving process. The basic selection takes place in two separate parts of the machine. In the first, the warp threads are selected so as to be either in the top shed, or in the bottom shed, so that after the weft has been passed through the shed, the warp threads will appear either on the face or on the back of the fabric. The warps can be selected individually, as in jacquard weaving, or combined into groups, the members of each group being selected together. The individual members of a group are held in the wires of a single heald shaft, and the heald shafts themselves are then selected so that each group moves into the top or the bottom shed.

The second selection is applied to the weft to obtain either different coloured wefts or different kinds of yarn in each pick. In conventional weaving this implies that the shuttle containing the required yarn is selected, and placed in position to be propelled across the reed. In shuttleless weaving it is necessary to select one of several small arms or fingers. These hold the end of the required yarn, which, when selected, is moved into a position where the gripper shuttle, rapier, etc. can pick it up and propel the yarn across the race. A similar process is used for selecting yarn in knitting machines.

The problems encountered in designing such mechanisms are of two basic kinds:

1 How to store the information as to which machine element, whether it be

heald shaft, shuttle, or knitting needle, is to be selected in any particular machine cycle.

2 How to transfer this information into the actual movement of the machine element. It is important that these movements are completed in a very short time, to allow the normal cycle of operation of the machine to continue without interruption.

6.1   METHODS OF STORING INFORMATION

The problems involved in storing the information can be separated into two categories, depending on whether the machine elements are to be selected to be in one of only two positions, or whether they are to be placed in one of more than two positions. Thus a heald shaft need only be selected to be in the up or down position, whereas the shuttle boxes must be selected to be in one of up to six positions, depending upon the number of different kinds of yarn that are being used for the weft. The most common selection requirement is the first. This

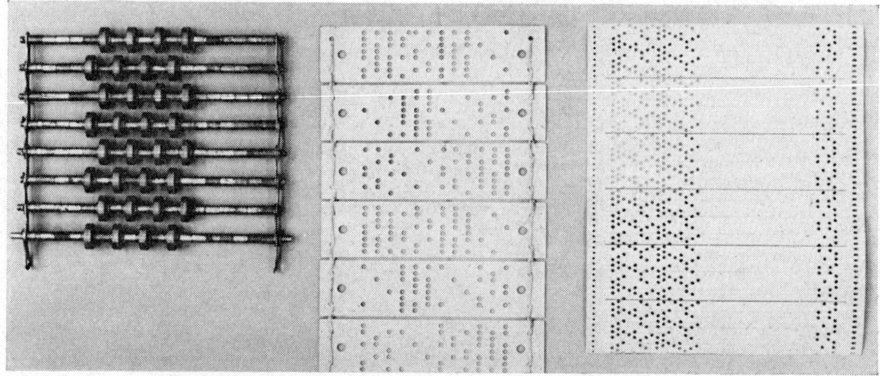

Fig. 6.1. Various forms of information storage.

can be considered to be an on-off requirement, regarding the up position of the heald as the on position, for example, and the down position as the off.

Information for such selection requirements can readily be stored in a form suitable for operating a mechanism, since the information store can consist simply of a set of elements, on each of which the presence or absence of some factor will initiate or prevent some mechanical action. These factors can consist of holes in the elements or of studs, bowls, etc. screwed or attached in some way to the elements. Fig. 6.1 shows a selection of several information stores used in textile machinery. They are: (a) punched cards, (b) punched endless tapes, and (c) bowls or studs attached to links which are joined together to make an endless chain. These endless chains and tapes usually consist of a set of rows of elements joined together, rather than a set of single elements, each element in such a row operating one of the many mechanical parts that have to be selected at each machine cycle. When these have been selected, and the machine cycle completed, the row of elements is replaced by the next row. This process is repeated until all the required selections have been made. It is usual for the pattern in the fabric to be repeated after a number of machine cycles, and this is very easily achieved by having the tape or chain in an endless form so that the pattern is reproduced for each revolution of the whole tape or chain. The number of rows in the chain will therefore be the same as the number of machine cycles required to make one repeat of the pattern.

This simple method of storing the information is suitable for patterns in which the positions of the different machine parts change frequently. This is the case when selecting the healds in looms or the needles on flat-bed jacquard knitting machines. In fact, it is normally the case when selecting for the purpose of making patterns in cloth. In other types of knitting machines where selection is also required for the formation of various garment shapes, such a method of storing the information would be extremely wasteful since, under these conditions, the machine parts remain in the same position for many machine cycles. For example, in a half-hose machine the machine parts will remain in the same positions while the leg is being knitted, or while the foot is being knitted; it is only necessary to alter the selection when changing from leg to heel or from heel to foot, etc. To simplify storage in such cases, the information is divided between two chains. One of these contains information about the required positioning of the various machine parts when a new selection is required. If such a chain were to be moved forward after each machine cycle a half-hose machine, for example, would knit one course of welt, one course of rib top, one course of leg, and so on. It is obviously necessary, therefore, to provide a second chain which stores information on how many courses must elapse before the first chain is moved to its next position, to produce the requisite number of courses in the rib top, leg, etc. of the sock being made. This is usually achieved by positioning a stud on the second chain element at that cycle of the machine when the first chain is to be moved.

Fig. 6.2 shows a half-hose knitting machine, together with its two information stores. In this instance the first chain is replaced by a drum which has studs screwed to it. The second chain is shown on the right of Fig. 6.2. The studs on this chain activate the movement of the drum. It should be noted that every row in the second chain contains only one element while the number of rows is equal to the number of machine cycles required for completing one half-hose. The number of elements on each 'row' of the drum is, however, equal to the number of machine parts which must be controlled, the number of rows on the drum being equal to the number of changes of selection required. In addition it should be noted that this method of storing information introduces a new principle. Instead of the presence or absence of a stud in the second chain indicating the *position* of a machine part it indicates whether the part, in this case the drum, is to be *moved* or not. This principle of information storage will be considered again later in this chapter, since it will be shown that there are certain means for producing the required machine movements which specifically require this type of information storage.

We shall now show how these methods can be modified to produce the required movement of machine parts which must be selected to be in one of many positions. One simple solution of this problem is shown in Fig. 6.2. The knitting cams and other elements have to be selected to be in one of three or four positions. This is achieved by having the studs on the selection drum of varying heights. This principle is also used to position the guide bars in a warp knitting machine. There are usually six possible positions of these guide bars, and the position occupied is controlled by the varying heights of a series of curved links, which

FIG. 6.2. Drum and information store on a half-hose machine.

pass round the selection drum (see Fig. 6.3) and act directly on a roller attached to the guide bars. The guide bar and roller are pressed against the link by a spring. It can thus be seen that the information has been stored in a form which is akin to that of a cam. This method of information storage is only employed in situations where such a cam can be used to position the machine element directly, without the intervention of any other selection mechanism.

While this is basically a simple method of selection, it has one major disadvantage. To position the elements at high speeds by these means requires accurately ground cam links. These are rather expensive, and as a result the patterning chain for a relatively simple warp-knit pattern will cost several hundred pounds. For looms, where these cam links would have to be larger, and therefore still more costly, such a system of selection becomes prohibitively expensive.

The on-off system of selection has, however, so many advantages that other ways have been devised for using it to select machine parts required to be in one

FIG. 6.3. Curved link chain selection on warp knitting machine.

of many positions. One of the most elegant, and certainly the most widely used, of these is to combine several on-off systems in such a way that the permutations of their on and off positions represent the several required movements of the machine parts.

For example, a chain can be constructed so that studs or holes can be placed in a number of different places across the width of the chain. Thus, if we denote the absence of a hole or stud by 0, and its presence by 1, one permutation on a chain that can carry a row of 4 studs or holes across its width would be represented by 1010. Another permutation would be 0110, and so on. The 'numbers' 1010 and 0110 are examples of binary numbers, i.e. numbers based on units of 2 rather than units of 10. Numbers based on units of 10 are termed 'dinary' numbers. The following table shows a few dinary numbers and their equivalents in the binary system.

Table 6.1

Dinary Number	Equivalent Binary Number
0	000
1	001
2	010
3	011
4	100
5	101
6	110
7	111

From this table it can be seen that we can represent 8 positions (numbered 0 to 7) by means of a row of three elements, which have either studs or holes to represent the units.

In the same way, if we have to select a position for a four-box mechanism, a chain consisting of rows of two elements can be used. If the positions are numbered 0, 1, 2 and 3, the layout of the studs and holes would be as shown in Table 6.2.

Table 6.2

Position	Binary Number	Position of Studs (S = stud; NS = no stud)	
0	00	NS	NS
1	01	NS	S
2	10	S	NS
3	11	S	S

It will be seen later that when this system is used the stored information is acted upon by a separate on-off selection mechanism for each selection element (hole or stud), the effect of the various mechanisms being added in some way to obtain the desired final positioning of the machine element.

It follows therefore that to position a four-box mechanism two separate selection mechanisms will be required. If three elements, together with three selection mechanisms, are used, these will be able to choose between positions 0 to 7, and would therefore enable us to operate an eight-shuttle box mechanism.

The binary principle has also been used, in an extended form, to replace the chain and drum type of store used on garment-making knitting machines. The drum contains information, for example, about the number of needles across which the narrowing operation is required for a large number of rows of knitting, and the chain therefore contains a number of blank links equal to the number of rows of narrowing required. A studded link then appears on the chain to move the drum to its next position.

In the Binary Cottons Patent flat-bed knitting machine the information required is stored in a single information store. Fig. 6.4 shows the type of store used. It consists of a series of rows of positions running parallel to the axis of the drum. The various positions in a row can be filled either by a hole indicating

FIG. 6.4. Information store for Binary Cottons Patent machine
(by courtesy of S. A. Monk Ltd).

the on requirement or by a blank indicating the off requirement. These rows of positions, where either blanks or holes may appear, are divided into two major sections. The first section consists of the first six positions. The second section, which consists of the remaining positions, contains, in binary form, information about the settings of the various machine elements. This section, therefore, takes the place of the drum in the older machines. The first six positions contain, in binary form, information about the number of machine cycles (i.e. knitted courses in this case) for which the machine settings—dictated by the second section of the information drum—remain the same. The first section thus takes the place of the chain. Since six positions are available in the first section, it is possible to have up to $2^6(= 64)$ repeating cycles of the machine.

This method of information storage is particularly useful for this type of garment-knitting machine since on these machines the narrowing or widening actions previously mentioned must be capable of being set to act across one to six needle spaces. On earlier machines this information was stored on the drum by having 'cam' sections of six different heights which could be attached to the drum as required. The information for this narrowing or widening action can, however, be stored more readily using only three positions on the binary code drum in the present form of information store.

The binary code method of storing information has become widespread in other technologies, such as digital computers, punched card information retrieval, etc., and a considerable technology has been built up on the basis of it. The availability of such a technology makes it probable that other sophisticated forms of information storage, retrieval and actuation may be introduced to various sections of the textile industry in the near future. At present, however, most of the methods used are relatively simple and are based mainly on purely mechanical selection mechanisms.

In concluding this section let us compare the various physical methods of storing information. Three mechanical methods have already been noted. These are (*a*) cam-like stores (warp knitting machines), (*b*) chains with studs or bowls, and (*c*) punched card, punched paper or plastic rolls. There is a further method of storing information, which consists simply of painting the required design on a set of squares ruled on a sheet of paper or on a drum. If a square is painted out the machine element is acted on; if it is not, the machine element remains inactive. Such an information store requires a photo-electric mechanism to operate it, and this is discussed in Section 6.7.

The cost of storing information decreases rapidly as we move from cams, through chains, to paper tapes and simple painted designs, but the complexity of the mechanisms required to activate the machine parts from the stored information increases at the same time. The most suitable system for a given application will therefore depend largely on the amount of information that must be stored. As an example, the amount of information needed to produce a plain woven fabric is very small, while that required to produce a jacquard weave is extremely large. It is not surprising, therefore, that the information required for weaving a plain fabric is stored in a simple cam system, while that required for a jacquard weave is contained in a large set of punched cards, or on a painted design.

The bulkiness of these methods of storing information is also very different. Fig. 6.5a shows a normal dobby operated on by a chain with studs, while Fig. 6.5b shows a similar dobby, producing a similar complexity of pattern, but operated by paper tape. The paper tape is obviously much less bulky and it is feasible to store such tapes for future use. This is relatively uneconomic with chains, due to the large capital cost of these chains and their large storage space requirement.

(a)

(b)

FIG. 6.5. a) Standard dobby with chain, and b) paper tape dobby (by courtesy of Geo. Hattersley Ltd).

FIG. 6.6a. Punched card label loom Jacquard

Fig. 6.6a shows the information store of punched cards required to operate the Jacquard on a label loom, while Fig. 6.6b shows the very much smaller information store needed when using a painted design.

All these means of storing information, except the painted design, are suitable for operating a purely mechanical selection system. Such mechanical selection systems are almost universal because of their high speed of operation and relatively low cost. They consist in essence of two separate sections, one to present the information to the mechanism at the right time and in the correct sequence, while the second section converts this information into the required movement of the machine parts. Before considering these mechanisms, however, it is as well to look a little more carefully at the parts of the machine which are moved.

FIG. 6.6 b. Photo-electric scanner for label loom Jacquard
(by courtesy of Apparatus Factory Ltd.)

6.2 THE GROUPING OF MACHINE PARTS FOR SELECTION

It has been assumed so far that every machine part that requires selection is selected on its own. This is the case in jacquard weaving and flat-bed jacquard knitting where every thread or knitting needle is selected individually. We have already seen that in normal weaving the threads can be grouped so that each group of threads is contained in one heald shaft. These heald shafts are then selected individually but the threads in any one group are selected together. This grouping of the threads clearly reduces the complexity of the mechanism required, and is possible because the pattern being woven repeats several times across the width of the cloth. Similar reasoning applies to the method used to select needles on circular jacquard knitting machines. There may be as many as 2,000 needles on a jacquard machine, and if each of these needles were to be selected individually the information chain would be needlessly complex, because the pattern usually repeats after about 20 courses. There is a further complication when selecting on circular knitting machines in that the needles act individually, passing the point of selection one at a time. If the selection chain were to operate on each needle it would have to be racked forwards one element for each needle. Thus for any one course there would have to be 2,000 links in the chain, and the complete chain for a pattern repeat in the direction of the wales would have some 40,000 links.

To overcome this difficulty the selection information is compressed into a much more compact form by grouping the needles together. If the number of wales in a pattern is n the number of groups will be $\frac{1}{2}n + 1$, when the pattern is symmetrical, and n when it is not. Patterns are usually symmetrical, and Fig. 6.7 shows the method of grouping employed for a symmetrical pattern repeating after 18 wales. (The number of groups of needles is therefore $(\frac{1}{2} \times 18) + 1 = 10$ groups.) The needles in any one group have a butt in the same position along the needle shaft. Thus needles 1 and 19 belong to the same group and have a butt in

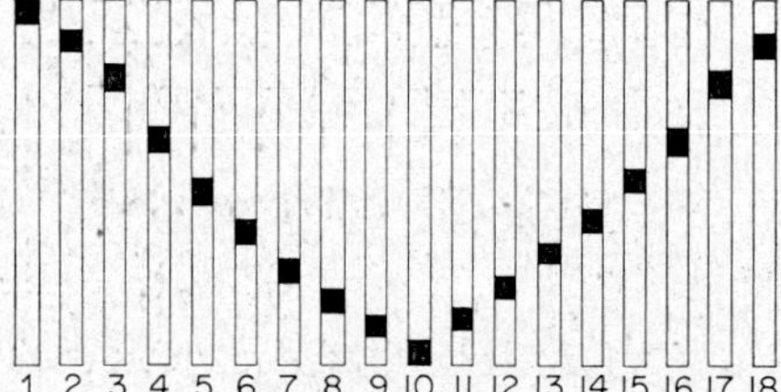

FIG. 6.7. Grouping of needles for an 18-wale repeat.

position 1, as shown. Similarly, needles 2 and 18 are in the same group, having butts in position 2, needles 3 and 17 have butts in the same position 3, and so on. This pattern of butt positions or grouping is continued round the circumference of the machine. Thus needles 2, 18, 20, 36, 38, 54, etc., are all in the same group.

All the needles with butts in position 1 are acted upon so that they perform the same action while the machine knits the first course. Similarly needles with butts in position 2 perform the same action, and so on. This selection is obtained by the use of a drum selector, a typical example of which is shown in Fig. 6.8. It can be seen that when this drum is stationary it presents one row of studs to a set of levers which act directly on the needles. This row of studs consists of a set of elements, in each of which a stud can be placed, or not, as required. If the stud is present the corresponding lever is raised, and every needle butt at that lever height is selected as it passes the selection point. There is a lever for every group height, and clearly, therefore, there must be as many elements in the row as there are groups of needles. The drum remains stationary while one course is knitted, but after each course has been completed the drum moves forwards one position and a new row of elements is presented to the levers, and a new pattern of selections made. This process is continued at each course until the pattern repeats itself down the cloth, when the drum will have made one complete turn. It can

Fig. 6.8. Brinton drum selector.

be seen that the number of rows of studs around the drum gives the number of courses over which the pattern repeats, while, as has been pointed out before, the number of elements in any one row is the same as the number of groups of needles.

This method of grouping the needles is very similar to the method used to group the threads in a warp in the heald shafts of a loom. It is usual to pass the first thread through an eye on the first heald, the second through the second heald, and so on, until sufficient heald shafts have been filled to make a half-repeat across the cloth (see Fig. 6.9). The subsequent yarns are threaded in the

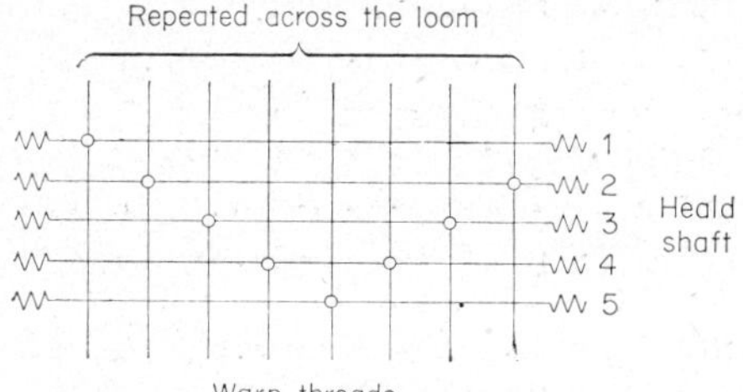

Fig. 6.9. Warp threading for repeating pattern.

same way as the needles are grouped (compare Figs. 6.7 and 6.9), producing the typical herring-bone type of grouping shown.

The method of grouping the needles or threads described above is only suitable for producing a symmetrical pattern, since a mirror image of the pattern given by the needles or threads in the first half of the herring-bone pattern will be produced by those in the second half. If the pattern is asymmetrical the needles or threads can be grouped in the way shown in Fig. 6.10. In this method the number of heald shafts or stud elements in a row is equal to the number of warp threads or wales, n, in a full repeat, instead of $\frac{1}{2}n + 1$. There are many other possible methods of grouping the needles or threads to produce particular requirements, but for further discussion of these the reader is referred to the many standard textbooks on knitted and woven structures. Suffice it to say, that

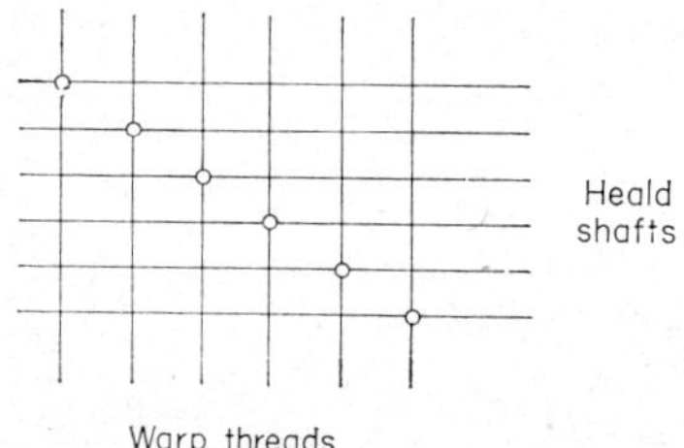

FIG. 6.10. Warp threading for non-repeating pattern.

the final pattern produced depends both on the selection chain used and on the method employed to group the threads or needles.

The mechanisms involved in selection divide themselves naturally into two groups. The first is concerned with transferring the information provided into a movement of a machine part, while the second is concerned with presenting the correct elements of the information store to this mechanism at the right time and place.

6.3 CONVERTING INFORMATION INTO MOVEMENT

The simplest means of converting the selection information into an up or down movement of a machine part is to make the machine part the follower of a cam which contains the selection information. The use of this system for warp knitting machines has been discussed in some detail in Section 6.1. The system is also employed in weaving, in cases where the number of heald shafts is relatively small and the pattern to be woven is relatively simple. Fig. 6.11 shows such a cam system for weaving plain fabric. Looms which select on this principle are known as tappet looms. It can be seen from Fig. 6.11 that if the cam shaft is rotated once for every two pick cycles the heald shafts will be raised and lowered at alternate cycles of the loom action, as required for a plain cloth. If a cloth is to be produced which repeats over four cycles of the loom the cam shaft will have to be driven at a speed such that it completes its rotation after four loom cycles. In addition, a different cam has to be used which has the required number

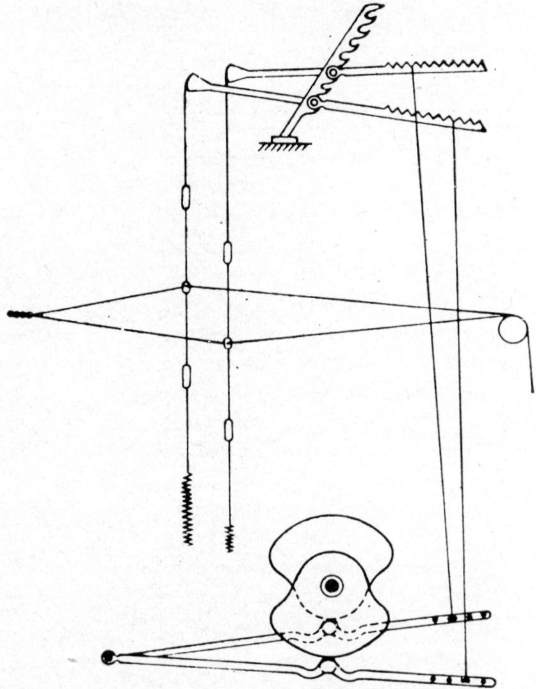

FIG. 6.11. Tappet selection on loom.

of lifts and dwells for the new pattern. This form of selection can be produced by negative cams, as shown in Fig. 6.11, or by employing positive grooved-disc cams of the type that were described in Chapter 4.

It is instructive to see why a chain of cam links like that used in warp knitting is not used for weaving. The distance moved by the warp knitting bar is relatively small, of the order of 0·03 to 0·1 in, while the heald shaft has to move a distance of some 6 in. It is possible, therefore, to use relatively small links in warp knitting bar selection without the intervention of any linkage to amplify the movement given by the cam. If a similarly sized link were to be used for weaving mechanisms the leverage in Fig. 6.11 would have to be such that the movement of the cam was amplified by a factor of about 20. However, this would result in the force on the cam itself being amplified. Since the heald shaft requires a force of up to 50 lbf to move it the result of the amplification would be a force on the cam of about half a ton. This very large force would be too great for a link of reasonable size to withstand. One can therefore see that the choice of selection mechanism depends to a large extent on the distance through which the machine part being selected has to be moved.

The tappet mechanism that we have just considered is clearly unsuitable when complex patterns are to be woven, or when frequent changes in pattern are

L

to be made. To give the machine greater versatility it is necessary to operate the selection mechanisms from the chain or punched paper information stores that were discussed earlier.

6.4 SOME MECHANICAL SWITCHING MECHANISMS

The mechanisms used to operate the machine parts are essentially switches. This can be seen most easily if we consider how the information on the store could be converted into a machine-part movement using, for example, an electrical system. In an electrical system any machine part that has to be selected could have a separate electrical motor connected to it, either through gears or by a linkage system. The holes or studs on the selection mechanism could then be made to close the contact on an electrical switch, so that the electric motor was switched on whenever the selector (hole or stud, for example) was in place. The motor would then move the machine part to its required position, a limit switch being included in the circuit to prevent the motor overrunning. When a hole or stud is absent from the selection mechanism a second switch is closed which operates the motor to return it to its original position, a limit switch again preventing the part from moving beyond its correct position. A similar system could be based on hydraulic motors.

It can thus be seen that the basic principle is to use a switch to connect the machine part to a source of power, which then moves the part. It is very important to realise that the selection element itself does not provide any power: it is merely used to activate the switch.

The system just described, using separate electrical or hydraulic motors for each machine part that has to be moved, would be much too expensive to use in practice, but the mechanisms that are used are, in fact, mechanical analogues of this system.

Before going on to describe them it is necessary to emphasise three functions that the mechanical system must be capable of fulfilling. These are as follows:

1 The machine parts must be connected to a source of power which will move them either up or down, or backwards or forwards. Thus if we denote the fact that the part must be in the up position by 1, and that it be in the down position by 0, a sequence of 1 0 1, for example, should make the part move up-down-up.

2 The movements, however, must be determined also by the sequence of required positions. Thus, if two successive machine cycles require the machine part to be in the same position, the mechanism must be such that the part does remain in position. Thus a sequence like 11 00 1 should produce movements of the part of the following sequence: up-stationary-down-stationary-up.

3 When a machine part is moved, it must move the required distance, no more and no less.

In the electrical system described above the first condition was fulfilled by having two switches, a forward and a reverse switch. The second and third conditions were both fulfilled by having two limiting switches. The types of solution used for mechanical analogues of this system are very different, and it is best to consider these by looking at specific examples.

The basic principles of weaving selection mechanisms are most readily understood by considering the Jacquard rather than the dobby. It has already been noted that the Jacquard selects each of many hundreds of threads individually, while the dobby selects one of, at the most, thirty-two shafts, each of which acts on a group of threads. The principle of selection in both cases is the same but, as

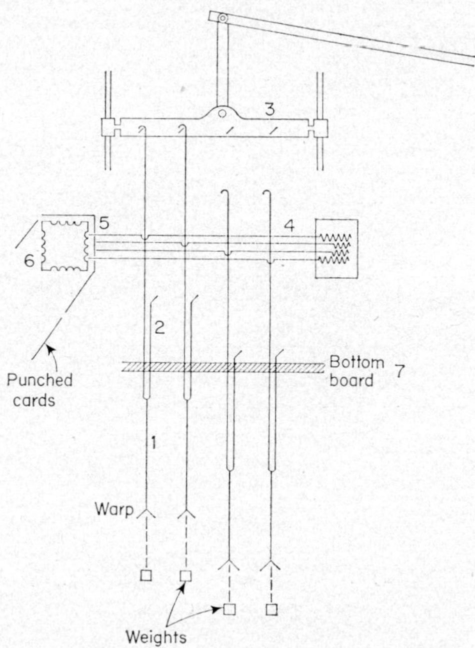

FIG. 6.12. Single cylinder, single-lift Jacquard.

jacquard mechanisms were the earlier invention, a consideration of the development of jacquard mechanisms leads us more readily to an understanding of the relatively complex, modern high-speed dobby.

Fig. 6.12 is a diagram of the simplest form of Jacquard, the single cylinder, single-lift Jacquard. In this mechanism each warp thread is attached through the cord 1, to a hooked member 2. The hook at the top of 2 can be engaged by a knife on the moving member 3, which is driven by a simple four-bar linkage mechanism so that it executes a simple harmonic motion in a vertical direction. When the hook is engaged, as the knife moves upwards, member 2 is raised, thus lifting the warp thread and opening the shed. The cord is weighted so that the hooks remain engaged with the knives during this motion. Whether or not the hook is engaged by the knife depends on the needle 4. The left-hand end of this needle (in the diagram) is pushed by a spring against a punched card 5, which is mounted on a square cylinder 6. If a hole is punched in the card opposite the needle, the needle will be pushed further to the left than if there was no hole in the card. In being pushed this far, the needle allows member 2 to move to the left in such a way that the hook on member 2 can be engaged by the knife, and the shed opened. If no hole is present in the card, member 2 is pushed to the right and the knife, on its upward travel, misses the hook, and the shed remains closed.

It is evident that when a hook is engaged by a knife it cannot be disengaged while the load is resting on the knife, nor would it be possible to re-engage the hook unless the knife moved below the hook position at some stage in its motion, so that the knife entered the hook from below. To achieve this a bottom board 7 is placed in such a position that the hook rests on it at a point in its downward path which is above the lowest point to which the knife descends. Fig. 6.13 shows the position of the knives plotted against time. The position below which the hooked member cannot travel due to the presence of the bottom board is also plotted on the graph. If a hook is engaged by a knife it will move down with it

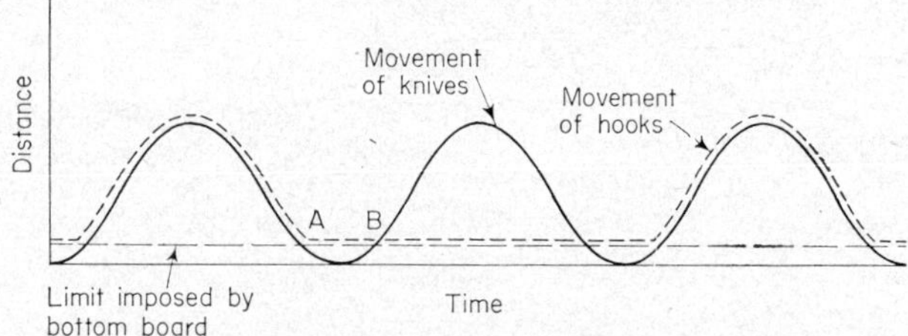

FIG. 6.13. Movement of hooks and knives in a single cylinder,
single-lift Jacquard.

but will leave the knife when point *A* on Fig. 6.13 is reached, and it cannot be re-engaged until the knife reaches point *B*. During the time the knife moves from *A* to *B* a new selection can be made, if desired. If the new card has a hole opposite the needle operating this hook, the hook will again rise with the knife after it passes point *B*. If a blank is present on the card the hook will not be engaged and will remain at the position dictated by the position of the bottom board. If a hook, in three cycles of the loom, is selected, not selected, and then selected again, i.e. the selection is 1 0 1, it will describe the movement shown by the dotted line on Fig. 6.14.

It should be noted that if a thread has been selected to be in the up position at two successive loom cycles, it will not stay up throughout the two cycles but will be lowered to its bottom position after each rise. It will be noticed, therefore, that while this mechanism fulfils the first and third conditions set out above, it does not fulfil the second condition precisely, since the threads are moved between two cycles in which it is required that they should be in the up position. This is clearly wasteful of power since the knives must raise the weights through their full possible traverse at each cycle. Since at the end of each cycle all the threads are down, and the shed is therefore completely closed, such a mechanism is known as a closed-shed Jacquard.

A development of this basic idea consists of a mechanism in which the board, as well as the knives, executes a simple harmonic motion. The two motions, however, are out of phase as shown in Fig. 6.14, in which the movements of the board and the knives are plotted against time.

To follow the operation of this system let us consider the movement of a hook that was initially in the up position (i.e. it was selected), but fails to be selected at the next machine cycle; on the next cycle, however, it is once more selected, i.e. a 1 0 1 selection. Initially the hook will be at 0 in Fig. 6.14. The knife then begins to fall and the board starts to rise. At point A, both are at the same height, and the hook is therefore disengaged from the knife. It then moves up again with the board until B is reached. At B the board is at its highest point and the knife at its lowest. Between B and C, the board moves down and the knife moves up, until they are once again at the same height (C). Since the hook is not selected on this cycle, it is not engaged by the knife at C, and therefore continues to move, first downwards, then upwards with the board until D is

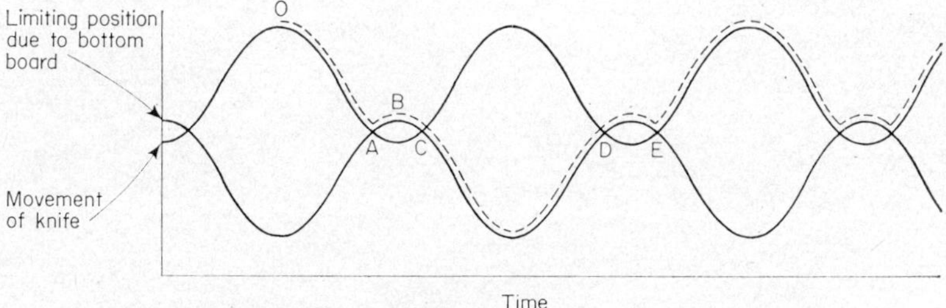

FIG. 6.14. Movement of hooks and knives in a single cylinder Jacquard with movable bottom board.

reached. The movement of the hook between D and E is similar to that between A and C. However, since the needle is selected on the next machine cycle, it will be engaged by the knife at E as the knife moves upwards. The path of the hook is thus shown by the dotted line in Fig. 6.14. Selection is made during the time intervals AC, DE, etc. It will be seen that, in this system, the hook always descends or ascends to its centre position, even if it is required to be in the same position at two successive cycles, and for this reason such a Jacquard is called a centre-shed Jacquard.

While some of the objections to the simplest Jacquard system have been overcome, it can be seen that the requirement of a stationary open shed has still not been met. A further development which uses two hooks and knives to operate each warp thread makes this possible. Fig. 6.15 is a diagram of one kind of double-lift Jacquard based on this principle. The two hooks 1 and 2 are attached through a set of cords to the warp threads in one of two ways. In the first method shown in Fig. 6.15, the two cords are simply attached to the thread. Clearly when one hook is raised but the other is not, one cord will be in tension and will raise the warp thread while the other cord will buckle, as shown, and be inactive.

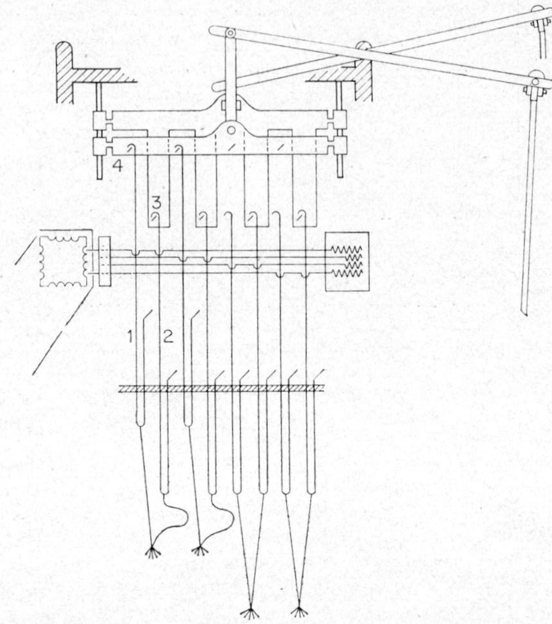

FIG. 6.15. Single cylinder, double-lift Jacquard.

The result of tying the warp thread in this way is to make the warp thread follow the path of the hook which is higher in its cycle. Fig. 6.16 shows an alternative method of attaching the warp thread to the two hooks. A cord passes round a

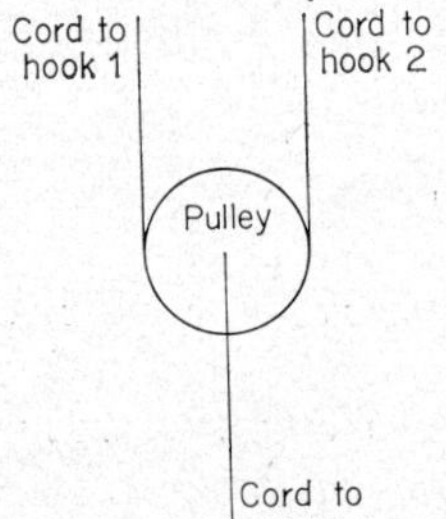

FIG. 6.16. Pulley method of averaging
two knife movements.

pulley and is attached to the two hooks. A second cord passes from the centre of the pulley to the warp thread, as shown. As a result the warp thread and the weight move through a path which is the average of the movement of the two hooks.

Whichever method of threading is used, the two sets of knives (3 and 4 in Fig. 6.15) are made to move up and down by two four-bar linkages, so that each performs a simple harmonic motion which repeats every two cycles of the loom.

The two motions are, however, out of step with each other, as shown in Fig. 6.17 which plots the movement of each knife against time. Also shown on this diagram is the limiting lower position of the two hooks, which is determined by the fixed position of the board.

It is not proposed to consider further the effect of the threading shown in Fig. 6.15 since it is fairly simple to show that this will produce a thread movement similar to that shown in Fig. 6.14, except that the lowered thread will remain down at successive cycles if it is not selected. This Jacquard is said to have a 'semi-open' action, and it is employed frequently. However, the threading shown in Fig. 6.16 is preferable, and moreover produces a fully open shed.

Fig. 6.17 shows the movement of the warp thread when the threading shown in Fig. 6.16 is employed, the dashed line showing the warp thread displacement for the following selection cycle. Both hooks are selected at the same time by the movement of the needle, but only one hook selection is effective because only one

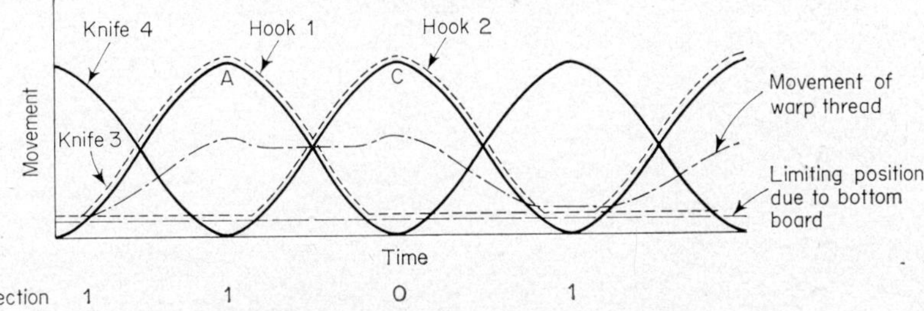

FIG. 6.17. Movement of hooks, knives and warp in open-shed
Jacquard.

of the hooks can be engaged by the rising knife 3. The other hook, which can only engage with knife 4, cannot be engaged because knife 4 is descending at this stage. By this selection, therefore, the warp is lifted until the hook reaches position A in Fig. 6.17. Knife 3 then begins to descend. During this cycle the position of the shed for the next cycle is chosen. In the sequence shown in Fig. 6.17 the shed is to be opened again. When selection takes place only one hook is resting on the board, namely, that which can engage with knife 4. Having been selected it is engaged by knife 4 on its upward travel, and consequently the shed is fully opened again at C. On the next cycle, the hook resting on the board is not selected and the other hook descends with knife 4 until it, too, rests on the board. The shed therefore remains closed. The selection cycle is therefore 1 1 0 et seq. The movement of the hooks is shown by the dotted lines in Fig. 6.17. The warp thread, whose displacement is the average of the two hook displacements, is also shown in Fig. 6.17 and it can be seen that the effect of this selection mechanism and threading arrangement is that the warp thread is maintained in its raised position throughout the first two cycles. This produces the required open-shed action.

This action is clearly desirable on mechanical grounds. It can be seen that the resulting action is smooth, and that no sudden impulsive forces act on the knives, such as occur in all other methods. It must, however, be noted that the thread movement is only half as great as the knife movement, and this may result in an undesirably large knife movement in some instances.

This brief survey of jacquard mechanisms is primarily meant to show that to obtain the most desirable mechanical action it is necessary to have two hooks and knives moving out of phase to obtain an open-shed action. The movement of the two hooks must also be averaged and then conveyed to the warp thread. This is the basic principle upon which most dobbies operate, as will be seen in the next section. For detailed consideration of the many refinements which have been incorporated into Jacquards the reader is referred to the selected reading at the end of this chapter.

6.5 DOBBY SELECTION MECHANISMS

It was shown in the previous section that to obtain the most desirable mechanical action an open-shed dobby must be used. Before considering such mechanisms it should be pointed out, however, that there are instances in which it is desirable from the weaving standpoint to have a centre-shed action. Attempts have been

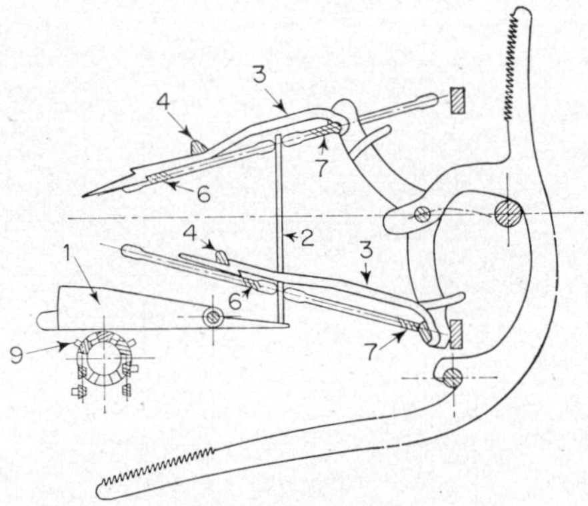

FIG. 6.18. Hattersley V-dobby (by courtesy of Geo. Hattersley Ltd).

made to produce centre-shed dobbies but, as there are no examples of such dobbies in production at present, they will not be considered further.

Fig. 6.18 shows the Hattersley V-dobby, on which most dobby actions are based. The primary elements of this dobby are the baulks, hooks, knives, pusher bars, etc., shown in Fig. 6.19. The knives move to and fro in a simple harmonic motion, each knife movement being out of phase with the other. The hooks are selected, according to the selection requirement, to engage with the knives when these approach the baulk. The knife and hook movement is identical with that shown in Fig. 6.17 for the double-lift Jacquard. Because the baulk is pivoted between the two hooks, and the heald shaft movement is taken from the centre of the baulk (see Fig. 6.18), the heald shaft movement is the average of the

movement of the two knives. In these respects the action is identical with that
of the double-lift Jacquard described in the previous section. It differs, however,
in two respects. Firstly the hooks return to their original position by the action
of the pusher bars shown in Fig. 6.18. These bars are rigidly attached to the
appropriate knives and move with them. This positive movement is necessary
as the heald shafts are not weighted and must be returned to their original
positions by the selection mechanism itself. Such a dobby is known as a positive
dobby, and is almost universal on modern looms. Due to the positive action of
the dobby it also requires two stopper bars, shown in Fig. 6.19. The force from
the warp on the baulk varies in direction depending on the position of the heald
shaft. It is consequently necessary to position the non-moving hook between two
stops. Note that, in the double-lift Jacquard, only a bottom board is required

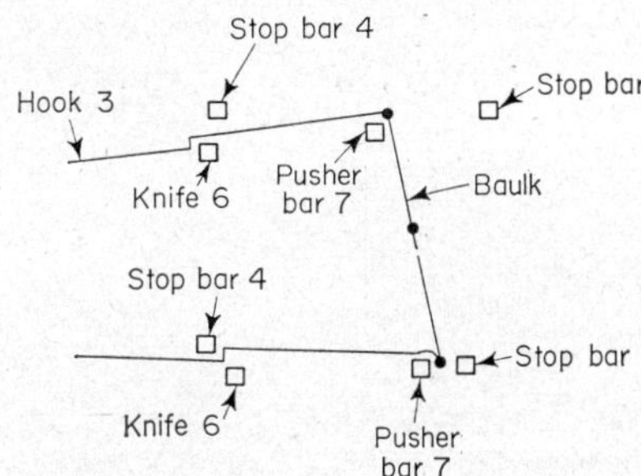

FIG. 6.19. Names of parts on V-dobby.

since the force on the cords is always in the same direction, due to the force
applied being mainly due to the weights.

The selection of hooks on the V-dobby takes place in a different fashion
to that of the double-lift Jacquard. In the Jacquard, only one selection needle
is used since the springy nature of the hooks makes it possible for the same needle
to select one hook, while the other is merely bent if it is engaged by the knife.
The hooks in the V-dobby are rigid, however, and it is therefore necessary
to have two separate needles to provide the selection. In Fig. 6.18 we see that
there is a lever 1, one end of which will be raised by the presence of a stud while
on the other end rests needle 2. There are two levers, one for each heald shaft,
one to operate the upper hook, and one the lower hook. Similarly, there are two
needles, one on each lever: a long one, as shown in Fig. 6.18, which operates
the top hook, and a short one, which cannot be seen, to operate the lower hook.
If a stud is present the lever tips up removing the support from the needle, which
then drops down. When the hook is in the inner position, as, for example, the
lower hook in Fig. 6.18, it rests on the needle. When the needle drops, the hook
drops under its own weight into the path of the knife. As a result, the hook will
be moved by the knife and pusher bar outwards and inwards respectively during
the next two loom cycles. It should be noted that this indication can be made
at almost any time during the last two loom cycles but will only become effective
when the knife has come to its inner position and can latch on to the hook. If no

stud is present the lever drops down under its own weight, thus raising the needle; if the hook is free of the knife it will be raised, and on the appropriate cycle the knife will pass the raised hook, which will therefore remain stationary during the next two loom cycles. If this 0 indication were made when the hook is on its inward return movement it would be raised before it was clear of the stopper bar 4; to prevent this happening the hooks are curved in such a way that they do not rest on the needle unless they are at their inner positions. Thus it can be seen from Fig. 6.18 that if needle 2 were raised the hook itself would not be moved until it had reached its inner position, when the downward curve of the hook would cause the hook to rest on the needle, so raising the hook under the influence of the lever weight acting through needle 2. Notice that on the inward movement, when the pusher bar is moving the hook, a clearance space exists between the knife and hook to prevent the hook fouling the knife as it is raised.

It will be seen from the above description that there must be two studs, or places for them, in each row of the selection chain for every heald shaft; the first

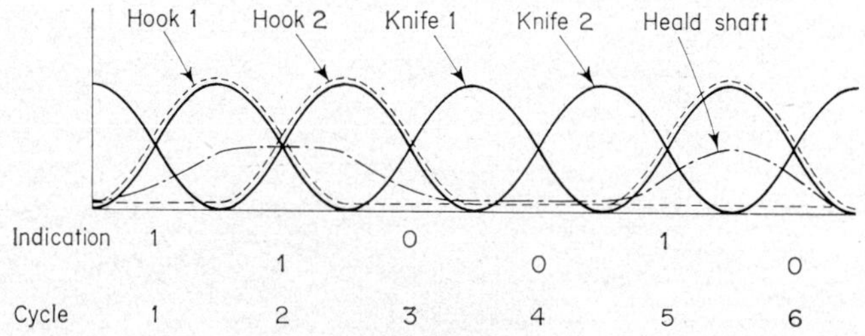

FIG. 6.20. Movement of hooks, knives and heald shaft (warp) in V-dobby.

for operating the upper hook, the second for operating the lower hook. Fig. 6.20 shows how a typical sequence of indications are arranged, and the resulting movement of the knives and heald shaft. The similarity to Fig. 6.17 is obvious and follows from the similarity between the action of the V-dobby and double-lift Jacquard. The two elements required for selections 1 and 2, 3 and 4, etc. are in the same row of the selection chain and provide the indication for the next two cycles of the loom, as the upper hook indication governs the heald shaft movement at cycles 1, 3, 5, etc. while the lower hook indication governs the heald shaft movement at cycles 2, 4, 6, etc. The information store is therefore moved forwards once for every two loom cycles. It can be seen from the above that this movement can take place almost any time during the loom cycle previous to the two loom cycles for which the indication is required.

We have thus far considered selection mechanisms which depend basically on dropping a hook into the path of a moving knife. It is possible, however, to achieve the same effect by bringing two gears into a position where they will mesh. Fig. 6.21 shows the essential features of the Knowles dobby, which is such a mechanism. The healds are connected to lever 5 (which is pivoted about a fixed centre E) in such a way that they are raised and lowered by the forward and backward rocking movement of lever 5. The mechanism, therefore, is essentially a means for producing this rocking movement of lever 5, at the right time and in the correct sequence.

Lever 5 is connected at D to another lever 4 which is pivoted at its other end B to an eccentric stud on gear wheel 3. This gear can be in mesh with either gear 1 or gear 2, the latter gears being driven by the machine in the directions shown. Consider for a moment what happens when gear 3 is in mesh with gear 1, as shown in Fig. 6.21. When this occurs gear 3 will be made to rotate in a clockwise direction. The point B will therefore also move in a clockwise direction and, after half a rotation of gear 3, will have moved over to the right, thus pulling lever 5 backwards. If only half a turn is made lever 5 will remain in this position, and the heald will be held in the raised position. If, now, gear 3 is connected with

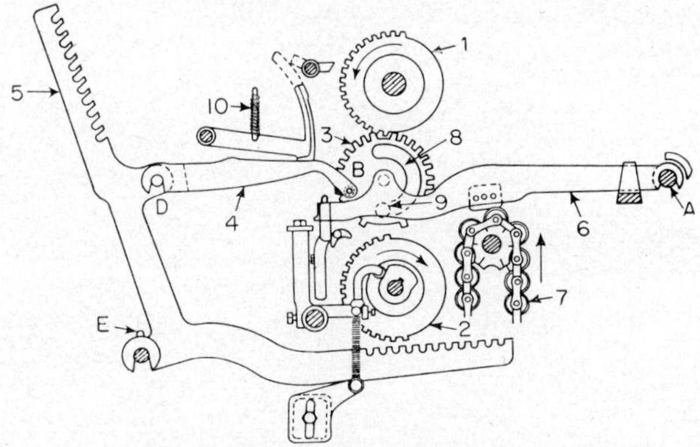

FIG. 6.21. Gear system in Knowles dobby (by courtesy of Crompton and Knowles Ltd).

gear 2, B will rotate in an anti-clockwise direction and lever 5 will be moved forwards. To prevent gear 3 rotating more than half a revolution whenever it is meshed with gear 1 or 2, these two gears have teeth over only half their circumference.

Which of the gears 1 and 2 is in mesh with gear 3 is governed by the control chain 7. It will be seen that gear 3 is centred on a lever 6, which in turn is pivoted about the fixed point A. When a stud is present on the control chain, it presses upwards on lever 6, thus engaging gear 3 with gear 1. If no stud is present lever 6 falls and gear 3 is engaged with gear 2.

The mechanism described so far satisfies the first and third of the three conditions, mentioned in Section 6.4, that a switching mechanism must satisfy, namely that the heald can be moved both up and down, and that the required movement must be always the same. The second condition, however—the provision for the possibility that successive machine cycles may require the healds to be in the same position—must also be satisfied. It should be noted that when gear 3 has been raised, gear 1 will raise the heald shaft if the next indication requires the heald shaft to be raised. Gear 1, which rotates once per loom cycle, will once more bring the geared half of the gear into contact with gear 3 and, if no provision was made to prevent this happening, would rotate gear 3 once more in a clockwise direction, and so lower the heald shaft. In the Knowles dobby this is prevented by having a gap in the teeth on gear 3. This gap, which can be seen in the lower half of gear 3 in Fig. 6.21, is four teeth in extent. This occupies about 30° of arc of the whole gear, and is large enough to prevent gear 1 or gear 2 meshing with gear 3 when this gap is reached.

To make this clearer, consider the sequence 1 1 0, i.e. two consecutive studs on the chain, followed by a blank. Initially gear 3 is engaged with gear 2. When the first stud presses against lever 6, gear 3 is raised and meshes with gear 1, thus being rotated so that the gap in its teeth is now at the top of gear 3. When the second stud presses against lever 6, gear 3 remains in the up position, but it is not engaged with gear 1 because of the gap in its teeth, and therefore gear 3 is not rotated and the healds remain in the position occupied in the previous machine cycle. In the next cycle of the sequence 1 1 0, no stud is present on the chain and therefore gear 3 drops onto gear 2. Since there is no longer a large gap at the bottom of gear 3, it meshes with gear 2, and lever 5 is thus moved backwards. It might be noticed that there is, in fact, a small gap in the teeth of gear 3, opposite the large gap. This is present simply to facilitate meshing between the teeth of gears 1 or 2 and gear 3 when a change of heald position is about to take place. The gap enables the first tooth of the half-gear circle to enter into mesh as gear 1 or 2 moves round towards the stationary gear 3.

There are three important locking devices on this dobby. Firstly, it has been shown that gear wheel 3 will only be rotated through one half-segment by either gear 1 or 2. However, at the completion of this motion gear wheel 3 could run on, due to its own inertia and the inertia and load of the heald shafts. To prevent this a semicircular slot 8 has been cut in gear 3 (see Fig. 6.21) and a fixed peg 9 resting in this slot prevents gear 3 from overrunning.

It is also necessary to hold lever 6 positively in place when the gears are meshing, as it will otherwise jump out of gear when meshing with gear 2, or place a large load on the stud when meshing with gear 1. To achieve this locking, a small mechanism is used, which is driven from the shaft of gear 2, and consists of a cam on the shaft which activates a lever having a knife on it. When lever 6 has moved into place then, immediately before meshing of the gears has occurred, the cam moves the lever and knife so that lever 6 becomes locked into position. After meshing the cam releases the knife, and lever 6 is free to occupy a new position.

Finally, it is necessary to prevent gear 3 from inadvertently rotating back on its own. A small spring 10 can be seen on Fig. 6.21, and this presses lever 5 downwards. Since it is necessary for lever 5 to move up when travelling from either the front to the back position or vice versa, this spring ensures that the fixed peg 9 always remains pressed against one end of the slot, except when gear 3 is positively driven by engagement with either gears 1 or 2.

The two mechanisms that have just been described are relatively complex because in both of them it is necessary to differentiate between, for example, an up selection which follows a down selection, and an up selection which follows a previous up selection. In the one case a movement of the heald is needed, whereas in the second case no movement is required. The switching mechanism could be greatly simplified if the selection chain were to store the information as to whether or not a *change* in position is required, rather than to store the information as to what the position itself is to be. Thus an up followed by an up, or a down followed by a down, can be recorded by a blank, while an up followed by a down, or vice versa, can be recorded as a stud on the chain. Thus the requirement that the heald positions should be 11 00 10 would be stored as 10 10 11. If this is achieved it is only necessary to ensure that the mechanism will always

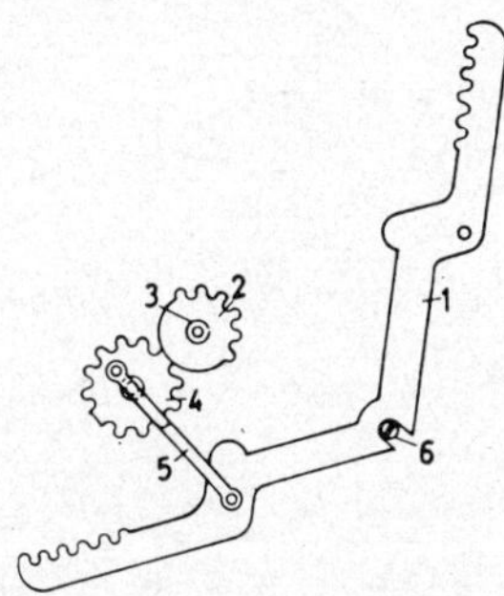

FIG. 6.22. Gear system in Leeming dobby.

reverse the heald position if the chain presents a stud to it. This has been accomplished very simply on the Leeming dobby, the essential parts of which are shown in Fig. 6.22.

The heald shaft 1 is pivoted about the fixed point 6 and is connected to a gear 4 by a lever 5. This lever is eccentrically pivoted on gear 4, as shown, and its other end is pivoted on the heald shaft. Thus, as gear 4 is rotated through one revolution the heald shaft will be raised and lowered. Gear 4 is made to rotate when it meshes with gear 2. This latter gear is driven by the machine, and rotates once during each machine cycle. It will be seen that it has teeth on only one half of its circumference; as a result, when gear 2 is meshed with gear 4, the latter is rotated through half a revolution while gear 2 makes a complete revolution. In this way, the heald shaft is alternately raised and lowered. Overrunning and chance movements of gear 4 are prevented by locking the concave surface of the outside of a tooth of double width on gear 4 with the convex surface of the non-geared portion of gear 2.

To provide for cases in which successive cycles of the machine require the heald shaft to remain in the same position, gear 2 is free to slide along its shaft 3. It is brought into the same plane as gear 4 by the presence of a stud on the selector chain, while if there is no stud present, it slides back out of mesh and no movement of the heald shaft takes place.

As an example consider the sequence 11 01, where 1 denotes the presence of a stud and o its absence. We will suppose that the heald shaft is initially in the up position. The first 1 in the sequence will cause gear 2 to mesh with gear 4, and the heald shaft will move to the down position. The second 1 in the sequence denotes the fact that another change in position of the heald shaft is required. Thus gear 2 remains in mesh with gear 4 and the heald shaft moves to the up position. The next number in the sequence is o, which indicates that no change in the position of the heald shaft is required. No stud is in the chain at this part of the sequence and gear 2 is therefore made to slide along its shaft and is disengaged from gear 4. Thus gear 4 does not rotate and the heald shaft remains in the up position. The final 1 in the sequence again denotes a change in position of the heald shaft, which as a result is moved to the down position.

Of the above three mechanisms, the last is by far the simplest, yet it is the most complicated type, the V-dobby, which is most commonly used for high-speed looms. To appreciate the reason for this, it is necessary to consider the detailed requirements of any mechanical switching action.

6.6 HIGH-SPEED MECHANICAL SWITCHING MECHANISMS

It can be seen from the above description of three typical mechanical switching mechanisms that they consist essentially of placing a stationary element, which we require to move, in the way of a continuously moving element. For example, in the Leeming dobby just described, the stationary element, consisting of gear 4 and the heald shaft, levers, etc., must be brought into contact with the moving gear 2. The force to do this is very small, since it consists only of the force needed to slide gear 2 along its shaft. But when the moving and stationary elements do come into contact the force between them can be extremely large because of the inertia of the element that has to be moved. The same could be said of any of the dobbies described above.

In the V-dobby the moving elements are the knives, the hooks being lowered into position where they catch the knives, to produce the large force needed to move the heald shaft. In the other two dobbies the moving and stationary elements are both gear wheels, a rotating gear tooth meeting a stationary tooth as the two gears move into mesh with each other. It is essential, if such an action is to be accomplished at high speed, that vibration and 'bouncing' of the two parts should not occur when the moving element meets the stationary element. This can most easily be accomplished by arranging the speed of the moving element in such a way that its speed is low when it meets the stationary element, but increases as soon as contact is established. Such a gentle initial action is difficult to achieve on switching mechanisms that use segmented gear wheels as the basic elements. It is for this reason that the gear dobbies have gradually become obsolete as loom speeds have risen and the need for greater care in bringing the elements together has inceased. The methods used for accomplishing such a gentle take-up in the switching action have thus become more complex in recent years as loom speeds have increased.

The earliest V-dobbies used the simple linkage system illustrated in Fig.

5.5a, which, as shown in Section 5.2, produces a roughly simple harmonic motion for the knives. The velocity of such a simple harmonic motion is zero at the two extremes of its travel, and if it were possible to engage the hooks and knives at the extremes of the knife movement, the take-up would be extremely smooth. To accomplish this, however, it would be necessary to have no clearance between hook and knife when the hook is dropped into place. The effect of even a fairly small clearance can be seen in Fig. 6.13 where the relatively small clearance provided by the bottom board on the Jacquard results in the knife striking the hook at point B at about half the maximum velocity. The clearance on a dobby is relatively smaller because of the geometry of the hooks and knives, but to obtain the required smooth take-up it would be necessary to have zero clearance. This is clearly impossible as some clearance space is necessary to ensure that the hook can drop cleanly into place at every machine cycle. The size of the clearance depends on two factors:

1 The accuracy with which the various components of the dobby have been made.

2 Whether the knife and hook always move in the same path or whether, because of wear in the machine bearings and slides, their paths vary slightly from cycle to cycle of the machine.

In the early dobbies the clearance required to ensure satisfactory running was fairly large, so that the knife at the extreme point of its travel was about $\frac{1}{8}$ inch away from contact with the hook. As a result the knife was moving fairly quickly when it engaged with the hook. To decrease this velocity the drive to the knives was later modified, so that it was taken from a cam drive instead of a simple linkage system. This modification brought about two improvements. Firstly, it was possible to make the contact between knife and hook more gentle, and secondly, it was possible to speed up the movement of the heald so that its movement occupied a smaller part of the loom cycle.

A more complex but more satisfactory means of achieving a gentle take-up in the contact between hook and knife is illustrated in Fig. 6.23 which shows the basic parts of the Staubli dobby.

The two knives in this dobby can rotate about their axes in addition to moving backwards and forwards in their slides. This rotation takes place as the knife has reached its inner position and during the period that it 'dwells' in this position. The knife movement is controlled by a cam to obtain this dwell. As the knife rotates it lifts the hook, as can be seen by comparing the upper and lower knives 1 and 2 in Fig. 6.23. The upper knife has rotated and has picked up the hook. Before moving out again the knife rotates back to its original position, and if lever 3 which takes the place of the selector needle in the Hattersley V-dobby, is moved out of the way the hook will drop with the rotation of the knife, eventually coming to rest with the knife fully engaged in the recess of the hook. When the knife begins its outward traverse there is therefore no shock loading on the hook. If lever 3 is not moved out of the way the rotation of the knife gently deposits the hook on lever 3, and when the knife begins its traverse it leaves the hook behind.

It should be noted that the selector lever 3 does not raise the hook, unlike the selector needles on the Hattersley V-dobby. Because the needle in the Hattersley V-dobby raises the hook it must be capable of providing a small, but not negligible, force to the hook and hence it follows that the studs on the selection chain must be capable of withstanding comparable forces. As a result the selection chain has to be constructed of substantial links and studs. This applies even more so to the Knowles dobby where even larger forces are required. The gentleness of the action of lever 3 makes it possible to consider operating the Staubli dobby, using the much cheaper paper tape as an information store. Because lever 3 has to move a considerable distance, however, paper tape has not

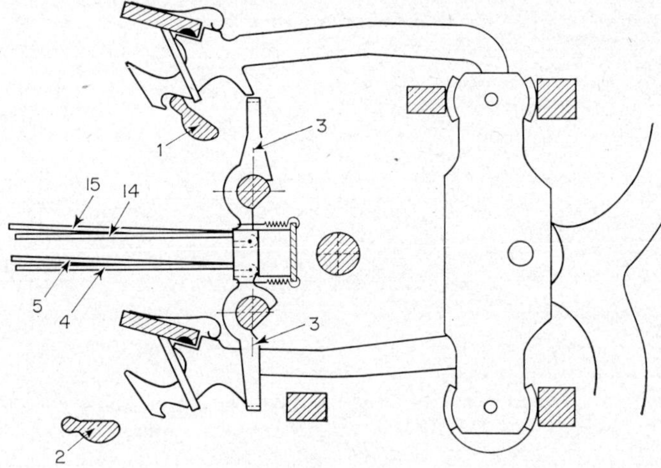

FIG. 6.23. The Staubli V-dobby (by courtesy of Staubli Bros and Co.).

been used to operate lever 3 directly but through a separate mechanical switching mechanism which operates lever 3 according to the information stored on the tape.

Fig. 6.24 shows the mechanism employed on the Staubli dobby. Other mechanisms are used on other makes of dobby, but it is universal practice to use such a two-stage action when operating dobbies from paper tape. The first stage operates a selector lever from the paper tape, this selector lever being capable of carrying only relatively small loads, while in the second stage the heald shafts are moved in accordance with the positions of the selector levers.

The basic principles of the Staubli dobby are the same as for the Hattersley V-dobby, there being two selector levers for each heald shaft, one to control the upper hook and one to control the lower hook. In the same way there are two positions on the paper card in each row for each heald shaft, one for the upper, and one for the lower hook. As in the V-dobby these two selections provide the information for two cycles, and the paper tape is only moved on to the next row after two loom cycles. There is, however, one extra requirement due to the mode of action of lever 3, Fig. 6.23. This lever must only be moved after the knives have rotated and have lifted the hooks clear of lever 3. The selection which has therefore been available throughout the whole of two cycles must only act on lever 3 at a particular time in one cycle, and this action takes place in different cycles for the levers acting on the upper and lower hooks. Fig. 6.24 shows the mechanism used to obtain this required selective selection of the two levers 3.

The selector levers are held by springs in their vertical position, when they will prevent the hooks from dropping. Two levers 4 and 5 are attached to the selector levers, lever 4 acting on the upper selector lever, and lever 5 on the lower.

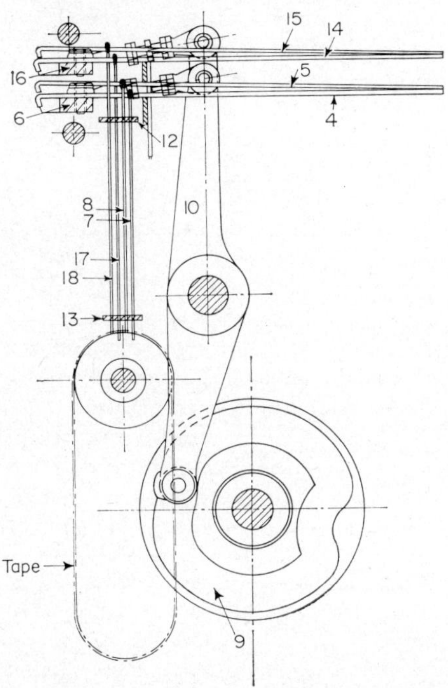

FIG. 6.24. The paper tape selector mechanism on the Staubli dobby
(by courtesy of Staubli Bros and Co.).

If either of the levers 4 or 5 are in the lowered position, as is lever 4 in Fig. 6.24, it will be pulled back by the sliding traction bar 6, and will therefore pull the selector lever 3 back, so that when the hook descends it will not be arrested in its movement but will become fastened in by the rotating knife and operated upon as described above. The position of levers 4 and 5 with respect to the traction bar 6 is determined by the needles 7 and 8 respectively. These needles rest on the paper tape, and if there is a blank in the tape opposite the needle it will remain in the up position, shown by needle 8, and the pull lever 5 will remain clear of the traction bar 6. If there is a hole in the paper tape the needle will drop through the hole (see needle 7), the lever will drop and will be caught by the traction bar at the appropriate time.

To ensure that traction bar 6 only operates on the pull levers at the appropriate time and in the appropriate loom cycle it has a series of grooves cut into it. A simple cam mechanism moves the traction bar in or out of the plane of the diagram in alternate loom cycles, so that when the upper knife is at its outer position the groove is opposite pull lever 4. When the traction bar operates, lever 4 is unaffected as it is opposite the groove, but at that stage pull lever 5 is not opposite the groove and, if it has been selected, the lever will be moved, so operating the selector lever. At the next loom cycle pull lever 5 will be in the groove and lever 4 will be acted upon, if it has been selected. To ensure that traction bar 6 moves the pull lever and the selection lever at the right part of the loom cycle, it is moved by means of cam 9 acting through lever 10.

Two other minor mechanisms are also shown in Fig. 6.24. These are required to prevent the fouling of the parts during certain movements. The first is required to lift the needles while the paper tape is moved onto its next position. The two needle bars 12 and 13 raise the needles clear of the paper tape between selection times, and then drop them when the paper tape has moved on to its next position.

The second mechanism is required to prevent the hooks from fouling the levers when the hooks are in their lowered position, and are being brought in from their outer to their inner positions. To prevent this occurring, lever 3 is moved out of the way immediately before the hook moves in, and is then returned to its original position ready for selection to be made by the pull levers when the hook has been raised by the knife. This movement of lever 3 is provided by levers 14 and 15 and traction bar 16. Traction bar 16 is moved by a separate cam not shown in Fig. 6.24. A complication of this system is that lever 3 must only be moved (*a*) when its hook has been selected and is therefore lowered, and (*b*) when this hook is returning inwards. This is necessary because a hook that has not been selected will be resting on lever 3 and must not be dropped in the path of the moving parts. The second requirement is met, as before, by having grooves in traction bar 16 so that levers 14 or 15 are only operated on when the knife is in its inward moving cycle. To prevent levers 14 and 15 from being acted upon when the hook is in the up position they are acted upon by needles 17 and 18, which are placed opposite the holes and blanks of the row in the paper tape previous to that making the present selection.

To see how this mechanism operates let us consider the upper knife to be moving inwards. If a hook has been selected in the last but one cycle it will now be in the lowered position and will tend to foul the lever. But since it *was* selected there must be a hole opposite needle 17. This needle will therefore drop and thus move lever 14 into the path of traction bar 16, so that no fouling of the knife by lever 3 takes place. If the hook was not selected in the last but one cycle it will be resting on lever 3, but now there must be a blank in the paper tape opposite needle 17; lever 14 will therefore be in the raised position and lever 3 will not move.

This relatively complicated first switching stage is made more complex by the use of lever 3 instead of the more usual selection needle, as on the Hattersley V-dobby. Simpler systems are available, but as in general such dobbies do not provide all the requirements of good mechanical switching they are of lesser interest. The two-stage switching action described above leads us naturally to consider the mixed electrical and mechanical selection systems which have appeared in very recent years.

6.7 MIXED MECHANO-ELECTRICAL SWITCHING MECHANISMS

It has already been mentioned that electrical, hydraulic and pneumatic systems are unsuitable for moving relatively heavy parts through fairly considerable distances at high speeds. The final stage in nearly all selection systems is therefore mechanical. It can, however, be seen that this objection does no apply to the earlier stages, where relatively small forces are needed to move the selection levers, needles, etc. A number of selectors have therefore been produced which work on the following principle. The information store consists of a sheet of paper on which squares may or may not have been inked, depending on the selection required. This has the great advantage that the store is in fact a 'picture' of the design required, and eliminates the necessity for making a more artificial information store. The sheet containing the design is passed stepwise under a photo-electric or infra-red scanning device. This consists of a beam of light or infra-red radiation, which shines on a row in the design. The information as to whether any square in this row is painted on or not is reflected back to a set of detector cells, one for each of the squares in the row. These cells then produce a small electrical signal which is amplified and used to raise or lower a small lever, hook or selection needle. This then selects a mechanical switch, which completes the selection action. For example, this method has been used to select the hook positions in Jacquards for label and name selvedge looms. Also, it has been used to replace the Brinton drum on circular knitting machines, the levers which the drum normally pushes in the way of the moving needles being operated on electrically by solenoids from such a photo-electric scanner.

This system is finding application in fields where it is either time-consuming to set up the normal mechanical store (the Brinton drum) or where the information store is frequently altered from one relatively complex design to another (the label and name selvedge looms). It has, however, also been used in an application where a mechanical switching action was unsuitable. The selection for a tufting machine requires the application of a braking force to a set of rollers situated in the feed creel. As these rollers are numerous and widely distributed the final switching is achieved by using pneumatic power to activate each brake, the pneumatic pressure being controlled by an electrically operated valve. The electrical signals for operating these valves are obtained from the infra-red scanning device described above.

6.8 ADDITIONAL COMPLEX MECHANICAL SWITCHES

The mechanical switches discussed so far have been concerned with converting a single on-off information code into a simple up-and-down movement. As has been previously mentioned, the same principles can be used to move a machine element to one of many positions. Box mechanisms are the classical examples of such complex switching systems, and Fig. 6.25 illustrates the basic principle of the Hattersley four-box mechanism. As the box must be moved to one of four positions, and as we have already seen, this will require two on-off selection elements on every row of the chain to indicate four positions. These two selection elements activate two Knowles dobby mechanisms similar to that illustrated in Fig. 6.21. The machine parts acted on by these mechanisms are not, however,

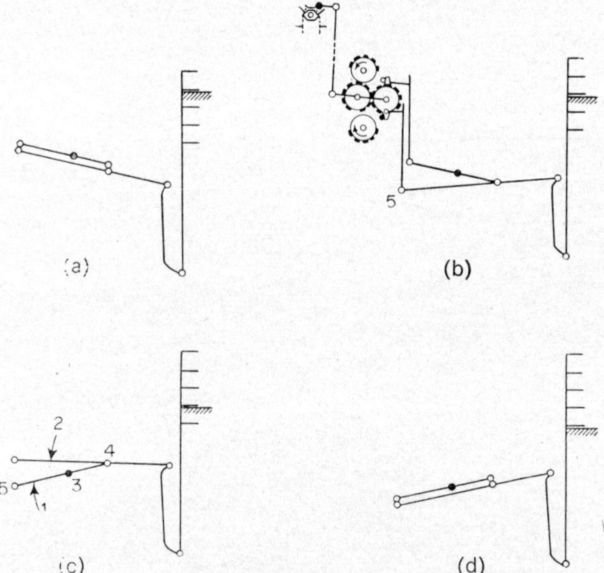

FIG. 6.25. Leverage system in a four-box mechanism.

heald shafts but the two levers 1 and 2 in Fig. 6.25. These two levers can therefore be in either an up or a down position and, to produce the required effect, moving lever 1 must result in the box moving through two spaces, while moving lever 2 must result in the box moving through one space. The movement of both levers 1 and 2 must also be additive. The linkage shown in Fig. 6.25 fulfils this requirement. Fig. 6.25a shows the box at its highest position with both levers in their lowest positions (the cross-hatched line shows the level of the race on the loom). The complete lever system pivots about the fixed point 3. If lever 1 is kept fixed then pivot 4 is also fixed and raising lever 2 results in the box dropping one space (see Fig. 6.25b). If lever 2 is kept fixed and lever 1 is raised, then pivot 4 will move down. As lever 2 is stationary pivot 5 is fixed, and therefore the movement of pivot 4 will be magnified at the box end of the lever. By the correct choice of lever lengths this results in the box moving down by two box positions, as shown in Fig. 6.25c. If both levers are raised then it is clear that the additional movement of lever 2 will move the box a further space, thus moving it three spaces in all, as in Fig. 6.25d.

It is possible in the same way to select one of eight box positions by using three Knowles dobby sections and the three-lever system shown in Fig. 6.26. This principle can be extended indefinitely but it is not used in practice for

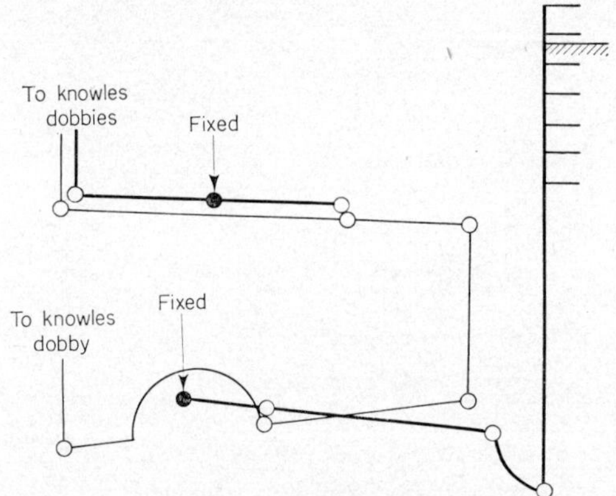

FIG. 6.26. Six-box motion leverage system.

selecting from more than one of eight positions. There are many alternative electrical and hydraulic methods for selecting from one of many positions (the telephone, for example, can select from one of many millions of phones) but these have not found any application in textile machinery because of the need in textile applications to provide large forces at relatively high speeds with equipment which is fairly inexpensive. As a result electrical and hydraulic systems, despite their inherent simplicity, have not been used

6.9 THE MOVEMENT OF THE INFORMATION STORE

In all the mechanical selection systems considered in this chapter one section of the store of selection information is presented to the switching system at some part of the machine cycle. Following this presentation, switching occurs in the mechanism. After this switching has occurred, and before switching occurs on the succeeding machine cycle, the next section of the store must be brought forward ready to be presented to the machine. On all the systems we have discussed the store must be moved forward through a fixed distance, then kept stationary until switching occurs. In chain and other systems the information is presented to the machine as soon as the store has moved forwards, but when using punched cards or paper tape it is also necessary to disengage the needles from the holes before the forward movement of the card or tape can take place. After the store has moved, this disengaging action must be reversed to present the information to the machine once more. On the Staubli dobby, for example, this was accomplished by moving the needles away from the paper, and returning them after the paper had moved, while on the Jacquard the card cylinder itself is moved back and then returned to the needles.

The method used for moving the information store, or for indexing the information it contains, depends to some extent on the size of the movement needed at each machine cycle, and some of these methods will now be described briefly.

6.9.1 Pawl-and-ratchet mechanisms

One mechanism that has been used successfully where small movements are required is the pawl-and-ratchet mechanism. For example, the Brinton drum on circular knitting machines is operated by such a mechanism. The pawl moves the ratchet forward once for every feeder point during each complete revolution of the machine. Selection is thus made for feeder 1, the next selection is made for feeder 2, then for feeder 3, and so on. Note that there are as many Brinton drums as there are feeders on the machine so that eventually every feeder produces the same pattern.

The pawl itself is operated by a cam consisting of a disc attached to the cylinder which has a series of rise-return sections on it, there being as many of these on the cylinder as there are feeders on the machine. This section is placed directly in front of the feeder and moves the Brinton drum on one space immediately before the next set of needle butts reaches it. To operate successfully, the drum must be prevented from slipping backwards between each of its forward movements. This is achieved by incorporating a separate, stationary pawl further back along the circumference of the ratchet wheel, and this ensures that the wheel can rotate only in the required direction.

6.9.2 Geneva mechanisms

The dobby mechanism requires relatively large movements of the store, and the forces required are also fairly large. The mechanism usually employed under these conditions is the Geneva mechanism, a typical example of which is shown in Fig. 6.27. The shaft 1 rotates continuously; thus the pin 2 describes a circle with constant angular velocity. When this pin reaches the position marked A in the diagram it engages in a slot cut in a wheel 4 mounted on shaft 3. Shaft 3 is thus rotated until pin 2 leaves the slot in the position marked B. The pin then continues to rotate but the wheel remains stationary until the pin again reaches A when it engages with the next slot on the wheel, and the process is repeated.

Fig. 6.27 shows a Geneva wheel with 4 slots, so that shaft 3 will make a quarter of a revolution for each revolution of shaft 1, this movement being completed in about a quarter of the time taken for shaft 1 to complete a revolution. During the remaining time, shaft 3 is stationary.

This mechanism therefore meets the necessary requirements of an indexing mechanism, but it is obviously desirable that the ratio of moving to stationary time should be capable of variation. This can be achieved by using different numbers of slots in the Geneva wheel, and between 3 and 16 slots have been used.

Another necessary refinement is some means of locking the Geneva wheel when it is not being rotated. This can be achieved by shaping the wheel between

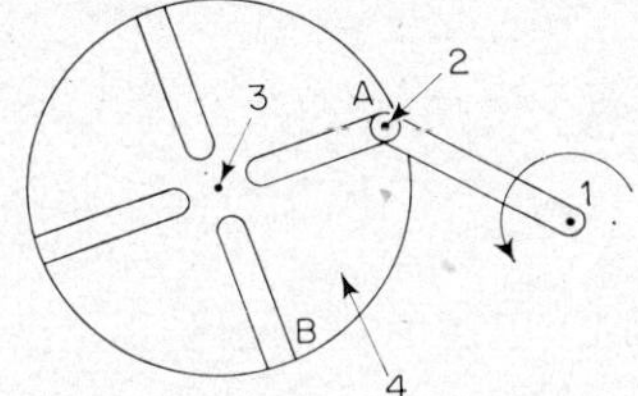

FIG. 6.27. Geneva mechanism.

the slots in the manner shown in Fig. 6.28, so that its edges are in contact with the cut-away wheel C when the pin is not engaged in the slot.

The Geneva wheel moves intermittently, continually starting and stopping as the pin engages and disengages from the slot. The motion of the wheel, however, does not involve any sudden changes in the angular velocity of the driven

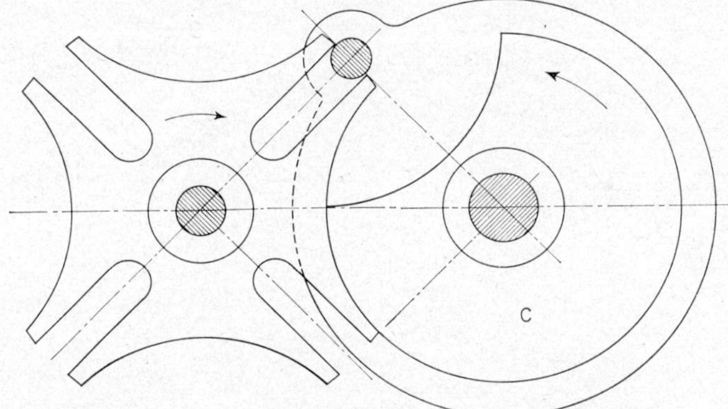

FIG. 6.28. Locking action on Geneva mechanism.

wheel. It can be shown that when the pin enters the slot, and thus begins to rotate the wheel, the angular velocity of the wheel is initially zero, but its angular acceleration is finite. A Geneva wheel is, in fact, an odd form of cam mechanism, and if we compare its motion with the cam systems described in Chapter 4, it can be seen that it has the same kind of initial behaviour as a parabolic cam. There is a sudden initial 'jerk' or pulse, but this does not involve an initial infinite acceleration such as is found, for example, in a linear cam.

Fig. 6.29 shows a variant of the Geneva mechanism used for paper dobbies, such as the Staubli dobby, in which the movements required for indexing are

smaller than those on conventional dobbies. This indexer can be regarded as a
multiple-slot Geneva mechanism in which, since the length of slot required is
extremely small, the wheel resembles a gear wheel. The 'tooth' shape, however,
is not that of a gear wheel, and is derived, in fact, from a series of short slots. The

FIG. 6.29. Indexer for paper dobbies.

number of slots used, however, leaves no room between the slots and makes it
impossible to use the same locking device as used on standard Geneva mechanisms. A separate linkage system is employed to move pin *A* when the wheel is
moved, then to replace it to lock the wheel when the driving pin *B* has completed
its indexing action.

6.9.3 Gear mechanisms

Another method sometimes used for indexing a dobby is shown in Fig. 6.30. This consists of a partially toothed gear, which is continuously driven, meshing with a complete gear wheel. The complete gear wheel is rotated by the segmented gear, but as soon as the latter has run out of mesh because of the missing teeth, the driven gear, i.e. the complete gear, remains stationary. The complete gear will be rotated at each cycle of the driven gear through an angle containing as many teeth as contained in the driving gear. In Fig. 6.30 the driving gear has 5 teeth, while the driven gear has 20. It will therefore require four rotations of the driving gear to obtain one complete rotation of the driven gear. The mechanism in Figs 6.27 and 6.30 will therefore reproduce the same indexing action. The indexing action can also be varied by changing the number of teeth on the

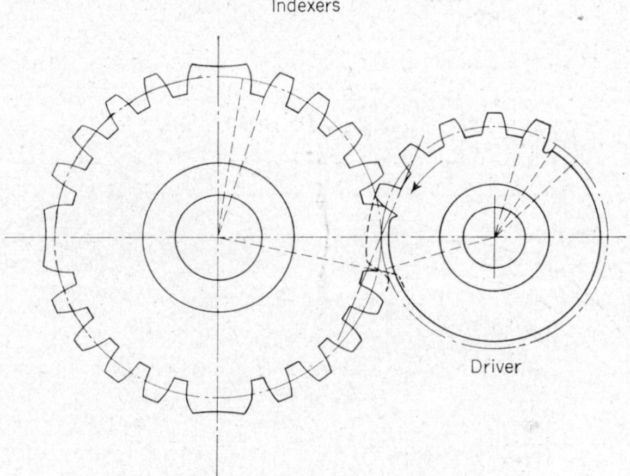

FIG. 6.30. Geared indexer.

segmented gear. Thus if only two teeth were used a movement similar to that given by a 10-slot Geneva wheel would be obtained.

In addition, the driven gear can be locked in position, when not driven, by the method shown in Fig. 6.30. Every fifth tooth on the driven gear has been enlarged to make a rounded section which locks into the curve of the non-geared section of the driving gear, so preventing any movement of the driven gear when not in mesh. The driven gear is really a 24-tooth gear with every fifth gear space filled in to make a double-sized tooth. This form of indexing requires the driven gear to move at the same surface speed as the driving gear in the instant that meshing occurs between the gears. An infinite acceleration is therefore required at the start of the motion, which can only be accommodated by the elastic deformation of the parts to smooth down the harshness of the action. This form of indexing is therefore not used on high-speed dobbies.

Fig. 6.31 shows a much smoother acting gear mechanism particularly suitable for paper dobbies where the movement required is very much smaller. It consists essentially of a worm wheel driven by the continuously rotating 'scroll' which is the worm. The pitch of the worm around its circumference is not even, however, and is arranged so that for most of its rotation the worm wheel is stationary because the pitch angle of the worm is zero. The pitch angle changes gradually until the worm wheel has moved forwards one tooth space when the cycle is repeated. The action produced by this indexing system is particularly smooth.

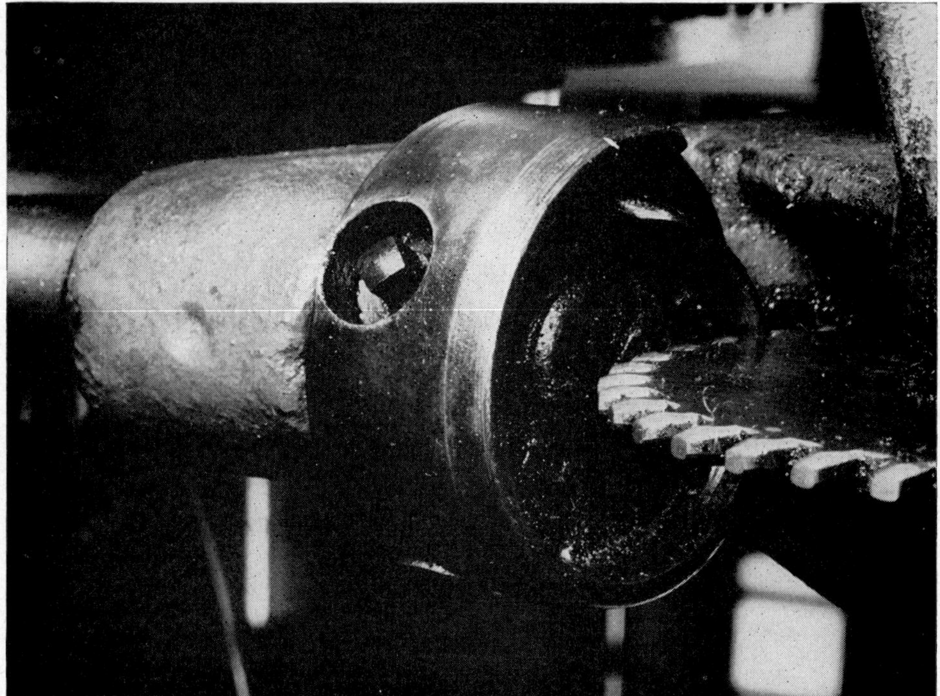

FIG. 6.31. Scroll indexer.

Indexing mechanisms are numerous, but the means used to disengage the needles from the cards are nearly all the same, consisting of cam-driven movements for moving the needles on paper dobbies (see Section 6.4), and linkage systems for moving the card cylinders on Jacquards.

This survey has been concerned only with the basic principles involved in the various selection mechanisms and the methods which can be used to analyse their behaviour. The following bibliography contains many detailed accounts of these mechanisms. They will be easier to follow if the basic principles that have been given here are kept in mind when reading such accounts.

Suggested Further Reading

T. Fox. *The Mechanism of Weaving*. Macmillan, 1932.

V. Duxbury and G. R. Wray. *Modern Developments in Weaving Machinery*. Columbine. 1962.

W. Middlebrook. *Loom Box Changing Motions*. Emmott, 1950.

F. Bradbury. *Jacquard Mechanisms and Harness Mounting*. F. King and Sons, 1912.

W. A. Hanton. *Automatic Weaving*. Ernest Benn, 1929.

R. W. Mills. *Fully Fashioned Garment Manufacture*. Cassell, 1965.

A. Reisfeld. *Warp Knit Engineering*. National Knitted Outerwear Association, 1966.

CHAPTER 7

CONTROL MECHANISMS

7.0 INTRODUCTION

The mechanisms described in the preceding chapters may often be considered to be means for performing mechanically operations that were formerly carried out manually by skilled operatives. There is, however, another aspect of an operative's job that must be considered, namely, that of taking decisions. For example, an operative in charge of a loom must make sure that the pirn in the shuttle is not empty, or nearly empty, when the shuttle is fired. If he decides that there is insufficient yarn on the pirn he must replace it by a new pirn containing yarn of the same type and colour. The whole of this supervisory task, which involves making a decision as to whether the pirn is empty or not, can be performed mechanically, and the mechanism used is an example of a class of mechanisms known as control mechanisms. Other examples of control mechanisms are the stop motions which stop a loom or a knitting machine when a thread breaks. In these cases, the mechanism decides whether or not yarn has broken, and takes the appropriate action of stopping the machine when necessary. It does not, however, replace the whole of the operative's task since it does not usually mend or replace the broken thread.

The examples of control mechanisms mentioned so far are of the type that operate intermittently, at widely spaced intervals of time. There are situations, however, in which the mechanism must act continuously. An example of this kind of mechanism is the autoleveller mechanism on a draw frame, which continuously controls the drafting ratio on the basis of a continuous measurement of the thickness of the input sliver.

The two types of control mechanism, the continuous and the discontinuous, are based on the same fundamental principles. There are, however, two essential differences between the two types:

1   The effect of the time delay that occurs between the mechanism making a decision and acting on it differs between the two types.

2   The response of a mechanism of the discontinuous type is nearly always the same. (A stop motion, for example, always performs the same action of stopping the machine when necessary.) In the continuous type of mechanism, on the other hand, the response of the mechanism must suit the requirements indicated by the input signal. For example, in the autoleveller, the size of the change in draft that is made depends entirely on the irregularity of the input sliver detected by the mechanism.

These differences will be considered in more detail later, but first we shall consider the basic elements of any control mechanism.

7.1 THE ELEMENTS OF A CONTROL MECHANISM

There are three basic parts to any control mechanism. These are: (*a*) a detector element; (*b*) a decision-taking element, and (*c*) an action element, which operates according to the decision taken by (*b*).

The decision-taking element can sometimes be so simple that it may be regarded as unimportant, but it must be emphasised that it is an essential part of a control mechanism. As an example, yarn-breakage detectors have only to decide whether or not the yarn is broken, but the fact that the decision is made is of basic importance.

The order in which the elements of the mechanism act need not always be that implied in the above list, namely:

detection—decision—action

Sometimes the detection can take place after the action. To appreciate this, consider how an operative would manually control the speed of a steam turbine. He would first operate a steam valve so that the speed of the motor would change. He would then detect the result of his action by reading the speed of the motor on a tachometer, and would then decide whether the speed was within the desired control limits. Having made his decision a further action takes place, and the sequence of operations would be:

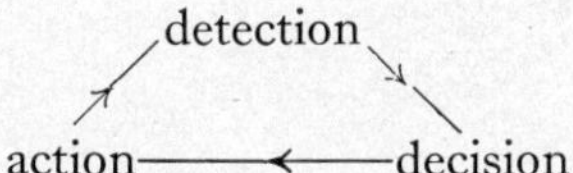

A control mechanism can be designed to operate in this sequence.

Thus control mechanisms can be divided into two types, according to their sequence of operation. Control mechanisms in which the sequence is in a loop as above are called *closed-loop* or *feed-back* mechanisms, and have been defined, rather aptly, as control devices that 'see' the results of their own actions.

All the mechanisms mentioned in the preceding section are of the other type, in which the sequence is detection-decision-action. These are called *open-loop* or *feed-forward* mechanisms.

How these systems are applied in general can be appreciated from the diagrams of the two types of mechanism shown in Fig. 7.1. The first shows a block diagram of a closed-loop mechanism. The control mechanism in this diagram consists of two parts, the detector and the controller, connected to the plant, or process, as shown.

When a disturbance occurs in the plant (a change in plant conditions) this is detected by the detector which sends a signal to the controller. The controller itself does two jobs. It first of all compares the signal from the detector with a reference signal, the magnitude of which is determined by the correct or desired condition of the plant. The difference between the two signals is then converted into an action by the controller, thus altering the plant in some way to compensate for the effect of this disturbance. This action is, of course, itself a disturbance which is detected by the detector and fed back to the controller. Thus, the controller (as mentioned above) 'sees' the result of its own action. From the direction of the arrows in Fig. 7.1a it can be seen that the loop connecting plant, detector and controller is closed, i.e. whatever the point at which one starts, one can, by following the arrows along the loop, return to that point. This loop does not consist of energy, matter or any other physical quantity flowing round

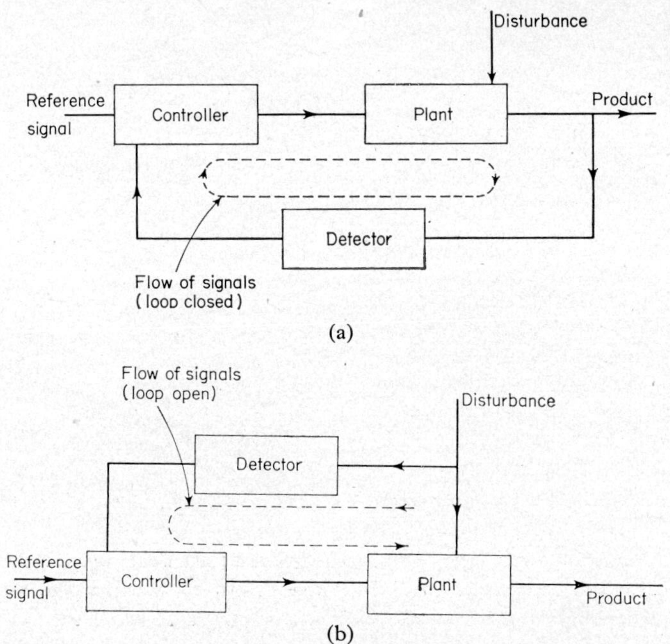

Fig. 7.1. Diagrammatic representation of open- and closed-loop systems.

the mechanism but what can best be called information which is continually moving round the loop.

The flow of information in an open-loop system, a block diagram of which is shown in Fig. 7.1b, is very different, though, as before, the disturbance affecting the plant is detected by the detector, and a signal sent to the controller. The controller again performs the same two jobs, comparing the detector signal with a reference signal and initiating action if necessary. However, there is no feed back, because the effect of the controller's action on the plant cannot be detected by the detector. In other words, the loop connecting the detector, controller and plant is open, i.e. starting, say, at the detector, and following the arrows round the loop, one finds, on reaching the plant, that the arrow representing the disturbance entering the plant prevents completion of the circuit. In fact the effect of the disturbance and the controlling action are added, and if the controller action is made correctly they should cancel each other out.

Nothing has yet been said about the kind of action that can be taken by the controller. This can be of two kinds. Either the action can have a magnitude that is a function of the difference between the detector signal and the reference signal, or it can be of an on-off character, i.e. if the difference is greater than a certain fixed amount the action is switched on, otherwise it is off.

Most of the control mechanisms that we shall be concerned with initiate actions of this second kind, whether they be open- or closed-loop systems. A simple example of an on-off action in an open-loop controller is a stop mechanism actuated by a yarn break. The purpose of the controller is to ensure that yarn is always passing to the machine. The presence of the yarn is often detected by means of a weighted arm resting on it. The position of this arm is compared with that of a fixed point below the yarn. If the arm drops below this point, the controller assumes that the yarn has broken and switches off the machine. In this particular case, as has previously been mentioned, the control action has normally to be completed by an operative who repairs the break and restarts the machine.

An example of the same type of action in a closed-loop system is provided by a normal bath thermostat. Here the temperature of the bath is detected or measured, and compared with the required temperature. If the bath temperature is too low, the heater is switched on. The effect of heating the bath in this way is detected and fed back to the controller until the temperature is too high, when the heater is switched off.

7.2 A COMPARISON OF OPEN- AND CLOSED-LOOP SYSTEMS

Two of the most important factors in the design of control mechanisms are the way in which the plant responds to the action initiated by the controller, and the time taken to respond. As an example of the latter, the stop motion of a knitting machine must be designed in such a way that it is capable of stopping the machine before the broken end of the yarn reaches the knitting needles.

The way in which the plant responds to the action is most important in closed-loop systems. Let us return for a moment to the operative who is manually controlling the speed of a steam turbine. Because this is a closed-loop system, the way in which the machine responds will affect the control system. This can be seen in the following way. Suppose that the machine is fairly slow to respond to the action of the operator (or controller). If the operator decides that the machine is running too slowly, and is also over-energetic in increasing the speed, then the speed will continue to increase for some time after the operative has noticed that he has increased the speed excessively. If he then decreases the speed too much, the system will begin to oscillate wildly about the required speed. This is a feature of closed-loop systems and is known as hunting. When hunting occurs the control mechanism is aggravating, rather than improving, the behaviour of the machine.

To prevent hunting the decision-making section of the controller must be very carefully designed, and the basis of such good design lies in the theory of automatic control, which is outside the scope of this survey. A reference to a simple account of the theory is, however, included in the bibliography at the end of this chapter.

There are four deductions that can be made from this theory that will be of considerable importance in our later discussion, and they are now stated without proof.

1   Hunting is more likely to occur if there are lengthy time delays in the system, i.e. if the plant is slow to respond to the action of the controller.

2   The effect of any delay in detection can be extremely severe. For example, if we are attempting to maintain a constant temperature in an air stream the temperature detector must be as close as possible to the heating device. If it takes some time for the air to reach the thermometer, the system will almost certainly hunt.

3   The difference between the detector signal and the reference signal is termed the error. The ratio of the correction made by the controller to the error is known as the *gain* of the system, and represents the 'vigour' with which the controller responds to any error detected. The greater the gain, the greater is the tendency of the system to hunt.

If the system is not hunting, however, then the accuracy of the controlling action can be increased by increasing the gain. This cannot be carried to extremes, though, because eventually the system will begin to hunt. Thus in all closed-loop systems there will be an optimum value for the gain.

4   An open-loop system cannot hunt.

Because of this last deduction, open-loop systems are much easier to design than closed-loop systems, and in some cases are the only systems that can be used. However, in such systems any error in the action applied by the controller cannot be detected, and hence allowance cannot be made. In a closed-loop system such errors can be accommodated, and this makes such systems much more versatile. In addition it should be noted that the open-loop system detects the disturbance, which presupposes that the cause of the disturbance is known. There are many cases in which so many causes of disturbance are present that they cannot all be detected, and in such cases only closed-loop systems can be used. These facts will be discussed further when specific examples of control mechanisms are being considered.

7.3   THE DETECTING ELEMENT

In the following sections we shall examine some of the various types of control mechanisms used in the textile industry. In doing so we shall find that the detecting elements will fall into a logical classification, and that in many cases the type of detector used tends to determine the rest of the control mechanism.

The detector can be basically a mechanical or an electrical device. Pneumatic and hydraulic detectors are also used, but very rarely in the textile industry.

Mechanical detectors are cheap and reliable but suffer from the difficulty of conveying the detected information over any considerable distance to the activating mechanism. In addition, the decision-taking section is usually much simpler and more compact if the information is conveyed to it in the form of an electrical signal. As a result, electrical detectors are becoming increasingly common in textile machinery.

The detector devices used are naturally very numerous because of the number of factors which it is desirable to control; thus we shall deal with them in the following sections under specific headings.

7.4 THE DETECTION OF BROKEN ENDS OF YARNS AND SLIVERS

One of the most common problems in textile processing is the detection of broken ends of yarns or slivers, and consequently the most frequently found type of detector is one that will do this job. Basically, all such detectors operate on the same principle. The yarn or sliver is passed through a small mechanical element in such a way that this element is kept clear of either a mechanical or electrical switching system. When the yarn or sliver breaks, the element moves (usually under gravity), closing the switch and thus operating the required mechanism.

An example of this type of detector is shown in Fig. 7.2, which illustrates

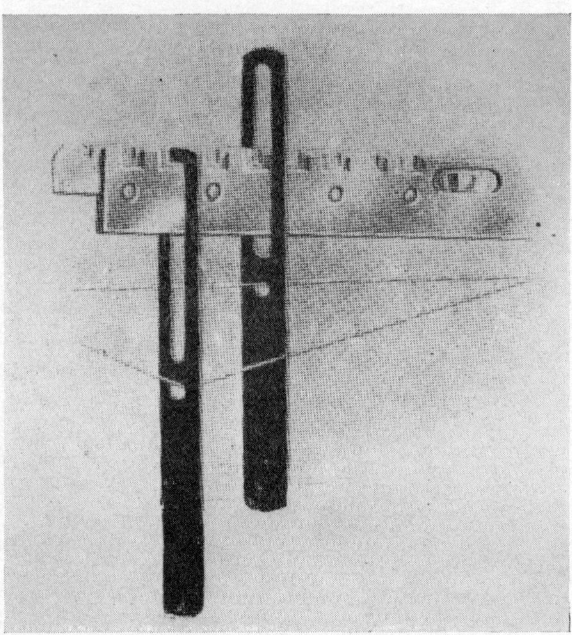

FIG. 7.2. Warp breakage detectors.

N

one method used to detect warp-thread breakages on a loom. Each warp yarn is threaded through one of the small metal parts, or drop wires as they are called, the tension in the yarns preventing the metal part from dropping into the serrations in the two bars shown in the illustration. The bars pass right across the loom, and the drop wires (one for each warp thread) are supported above the bars in this way. One of the bars oscillates continually, while the other is connected to the loom-stopping mechanism and is normally stationary. If a thread breaks, the drop wire, due to gravity, falls into one of the serrations, and as a result the stationary bar is pushed forward by the moving bar, and the loom stopped. In an alternative but similar system there is a pair of stationary bars which form the two ends of an electrical circuit. When the drop wire falls and touches both bars the circuit is completed and the machine stopped. These two examples show how mechanical and electrical switching can be used as alternative means for obtaining the same result.

Fig. 7.3, on the other hand, shows a detector which could not readily be operated mechanically. This detector is used to stop a circular knitting machine when the yarn being fed in breaks. The yarn tension keeps the arm A down so

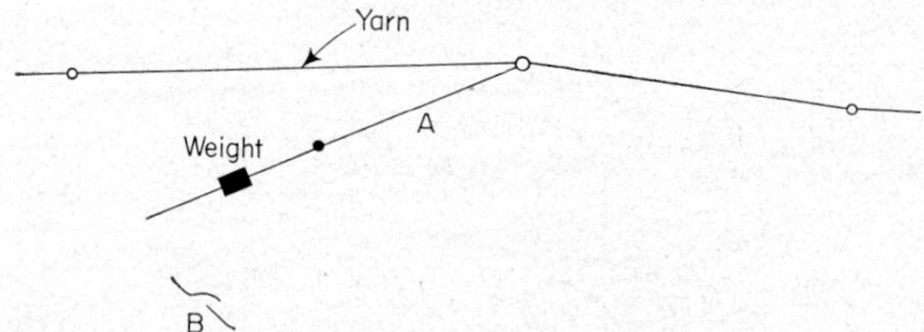

FIG. 7.3. Yarn breakage detector.

that the electrical switch B is not closed; as soon as the yarn breaks the arm flies up, the switch makes contact and the machine is stopped.

An interesting variation on these detection mechanisms is the method used to detect the absence of weft during the weaving cycle. The method is basically the same as those described above, but in this application the mechanism has to allow for the fact that during the part of the weaving cycle, when the weft is not being laid in to the shed, there is no weft for the mechanism to detect. In other words, the mechanism must be inactive for part of the machine cycle. An example of this kind of mechanism, known as a centre-weft fork mechanism, is shown in Fig. 7.4.

The mechanism is situated in the centre of the race, and the presence or absence of yarn during weft insertion is detected by the two weft prongs 7, which rest on the weft. If weft is present the prongs are supported by the weft and prevented from dropping into the recess in the race. When this is the case, the loom-stopping mechanism is not activated and weaving is uninterrupted.

If, however, the weft breaks (or for any reason is absent) the prongs fall into the recess in the race, and in so doing operate the mechanism that stops the loom. The prongs will, of course, drop into the recess after the weft is beaten in and the race moves towards back centre. At this stage, however, the stop mechanism is made inactive by the sley movement since the mechanism has already ensured that weft is present.

The forks must be raised above the level of the weft immediately before the shuttle is fired. This is accomplished by a small cam system operated by the sley, thus ensuring that the forks are correctly synchronised with the position of the reed. Fig. 7.4 shows that the mechanism required for accomplishing this is considerably more complex than the simple mechanism shown in Fig. 7.2, and this follows directly from the fact that the decision-taking section is so much more complicated. It is necessary to stop the loom if weft is absent at a certain period of the loom cycle but to ignore its absence or presence at all other periods.

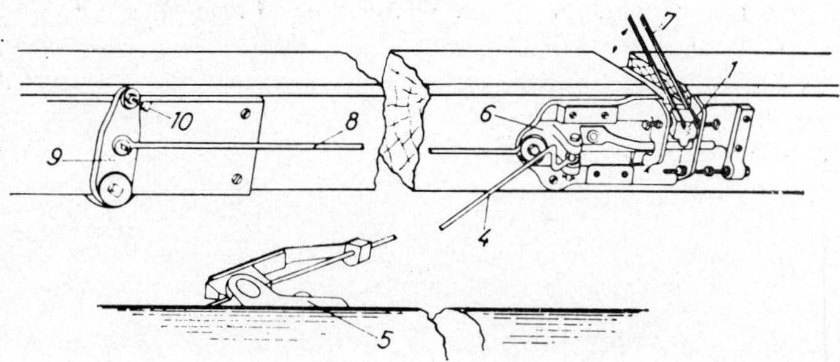

FIG. 7.4. Weft fork detector.

A completely different method of detecting the presence of a broken warp thread is that used for detecting broken ends in warp-knitting machines. The large number of threads, their relative closeness and delicacy, combined with the high speed of operation of the machine, make devices such as that shown in Fig. 7.2 impractical. The broken ends in warp-knitting machines are therefore detected by examining the cloth after knitting. This cloth is examined by passing a light source to and fro across the cloth close to the knitting bars. If a thread is broken a hole or 'ladder' appears in the cloth, and more light than usual passes through the cloth to a photo-electric cell moving in synchronism with the light source. An electronic device detects this increase in illumination, and causes the machine to be stopped whenever it occurs.

7.5 THE CONTROL OF YARN AND CLOTH TENSION

In almost all textile processes it is important that the yarn, sliver, or fabric should be maintained at a reasonably constant tension, since otherwise the properties of the resultant material will be affected. Consequently many mechanisms have been devised to control tension in yarns and cloths.

One of the most rudimentary methods used is not a true control mechanism. The basic idea of this method is to add a known tension to the yarn or cloth by means of a device usually called a tensioner, which can be of many different kinds. Fig. 7.5, for example, shows a simple let-off mechanism for applying tension to the warp on a loom. This mechanism consists of a simple rope brake, the rope being wound round the shaft of the warp beam. The amount of retarding torque applied to the beam can be varied by changing the position of the weight on the lever arm attached to one end of the rope. This form of let-off, in which the warp beam is not positively driven, is usually known as a negative let-off.

Fig. 7.6 shows two yarn tensioners, which again are essentially friction brakes. The amount of tension applied can be varied in these devices, either by changing the sideways pressure on the posts of the gate tensioner or by adding weights to the top of the disc tensioner.

The amount of tension added by these devices, however, is not truly constant because it depends entirely on the coefficient of friction between yarn and

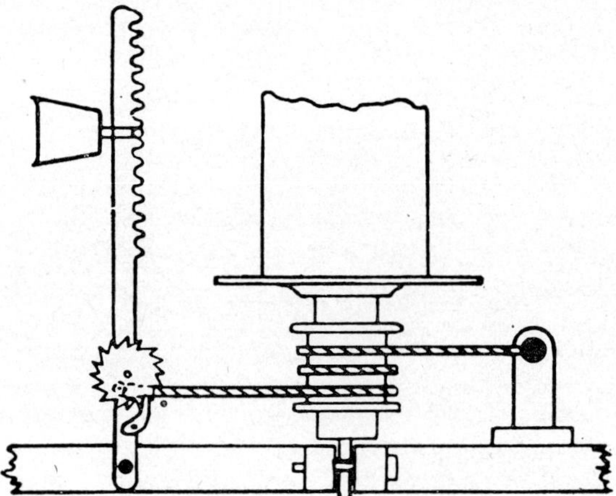

FIG. 7.5. Negative let-off tension control.

metal, in the case of the yarn tensioners, and between the rope and the warp beam in the case of the let-off mechanism. This friction coefficient can, due to a multiplicity of causes, vary with time and consequently the added tension varies. In addition, in the warp let-off mechanism the applied tension will vary as the diameter of the warp beam decreases during weaving. The reason for this is that the rope applies a constant torque to the beam shaft, which the warp tension must overcome. Hence as the beam diameter decreases, the warp tension will increase.

Such simple 'controls', therefore, have many drawbacks, and because, in addition, they cannot be used for cloth take-up tension control, it is usual nowadays to fit closed-loop control mechanisms except where the cost would be prohibitive as, for example, in warping.

A closed-loop system that has been used to control yarn tension is shown in

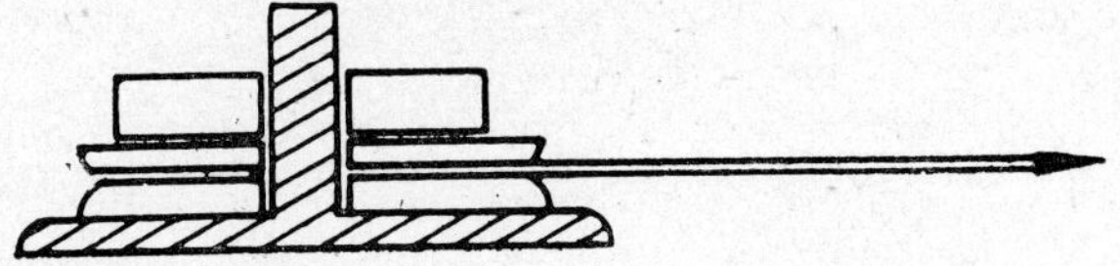

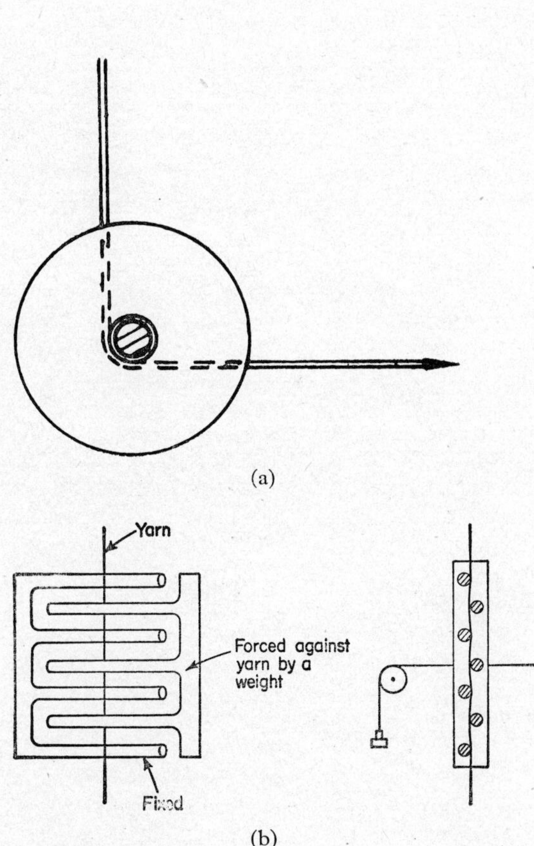

(a)

(b)

FIG. 7.6 a) Disc, and b) gate yarn tensioners.

Fig. 7.7. This consists of a spring-loaded foot F under which the yarn passes. Attached to this foot is a lever FAH with its fulcrum at A. The yarn passes through a hook at H, and the lever acts as the tension detector. The system works in the following way. Suppose T_o is the input tension to the device and T is the output tension which is detected by the detector lever. Then the downward tension T on the end H of the lever will produce an upward reaction at

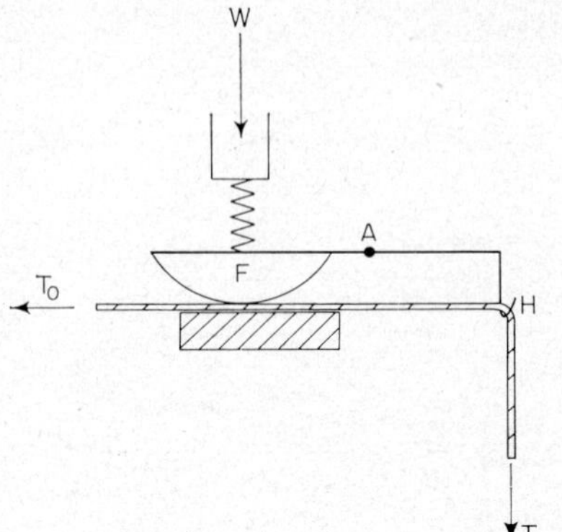

Fig. 7.7. Simple tension control.

the foot equal to αT, where α is the mechanical advantage of the lever (equal to the ratio of lengths AH/AF). Thus if W is the load on the foot provided by the spring, the effective downward load on the yarn passing under the foot is $W - \alpha T$. Hence if μ is the coefficient of friction between the yarn and the foot, the frictional resistance to the yarn is $\mu(W - \alpha T)$. This frictional resistance is added to the input tension to give the required output tension.

Suppose that T_c is the required output tension; then the frictional resistance would be $\mu(W - \alpha T_c)$. But the system should be arranged so that when the output tension is correct, no additional tension would be added to the yarn. In other words, the frictional resistance should be zero when $T = T_c$. Consequently the load W would be adjusted so that

$$\mu(W - \alpha T_c) = 0$$

or

$$W = \alpha T_c \tag{7.1}$$

The value of T_c is the reference signal of the device.

Now imagine that the output tension changes from T_c to T', i.e. there is an error in the system equal to $T_c - T'$. The frictional resistance (i.e. the added tension) now becomes $\mu(W - \alpha T')$. Thus the correction applied to the tension is equal to the change in frictional resistance, which is initially zero, when $T = T_c$. Hence the correction that has been applied is:

$$\mu(W - \alpha T') = \mu\alpha(T_c - T')$$

since, as we have seen, $W = \alpha T_c$. Consequently

$$\text{Gain} = \frac{\text{Correction}}{\text{Error}}$$

$$= \frac{\mu\alpha(T_c - T')}{T_c - T'}$$

$$= \mu\alpha$$

It is instructive to show that if the gain is large, the output tension will always be equal to the required tension T_c, however much the input tension may vary. We have already seen that when the output tension is T, the frictional resistance is $\mu(W - \alpha T)$. But, because the yarn is running at constant speed and therefore at zero acceleration, this frictional resistance must be balanced exactly by the difference between the input and output tensions, i.e.

$$T - T_o = \mu(W - \alpha T)$$

which gives:

$$T = \frac{T_o}{1 + \mu\alpha} + \frac{\mu\alpha}{1 + \mu\alpha}\left(\frac{W}{\alpha}\right)$$

From this equation it can be seen that if $\mu\alpha$ is very large, $T \simeq W/\alpha$. But W/α is equal to the reference tension T_c, from Equation (7.1), and therefore the output tension is always equal to the required tension.

The above formula shows an interesting fact, that the effect of the disturbance T_o is reduced by a factor ($1 +$ the gain) by the use of a closed-loop control. It can be shown that this is generally the case, and this has already been commented on in the introduction to this chapter. In addition, however, the effect of a high gain is to make the system liable to hunt, and such simple tension controls have required some damping to be fitted, to prevent violent oscillations. This, however, has resulted in the controlling action being sluggish.

7.6 LET-OFF AND TAKE-UP MECHANISMS

Cloth take-up and warp let-off mechanisms are usually described as tension controls, but the description is not apt since in no case do such mechanisms involve measurement of tension. The principle on which they operate is illustrated in Fig. 7.8a. Cloth or yarn is fed from the beam at a linear speed v_1, and is taken up by the second roller at a linear speed v_2. The cloth or yarn passes between the rollers, as shown, and over a weighted roller (Fig. 7.8a) or a spring-loaded roller (Fig. 7.8b), which takes up any slack in the yarn or cloth passing between the rollers. In a take-up mechanism v_1 is fixed by the rate at which yarn or cloth is fed to the machine. The control mechanism detects the position of the floating roller; if it is not in the reference position, the control system alters the speed of the take-up roller. Such a mechanism has been used on jig dyers and finishing machinery. On looms, however, it is the let-off which is controlled. The velocity v_2 is fixed by the rate at which yarn is used in making the cloth, and the control mechanism varies the speed v_1 to return the floating roller towards its controlled position. Clearly if the floating roller rises the take-up speed is greater than the let-off speed, while if it drops the opposite applies. In the example shown in Fig. 7.8a the effect of the weight on the floating roller is to maintain a constant tension on the cloth, but this is a by-product of the mechanism, and not a result of its control action.

Fig. 7.9 shows a typical loom let-off mechanism. The floating roller which determines the path length of the warp is the roller A. The weight is applied to the roller through the lever D. The position of the roller is conveyed to the lever C, and levers C and P are so arranged that the difference between the

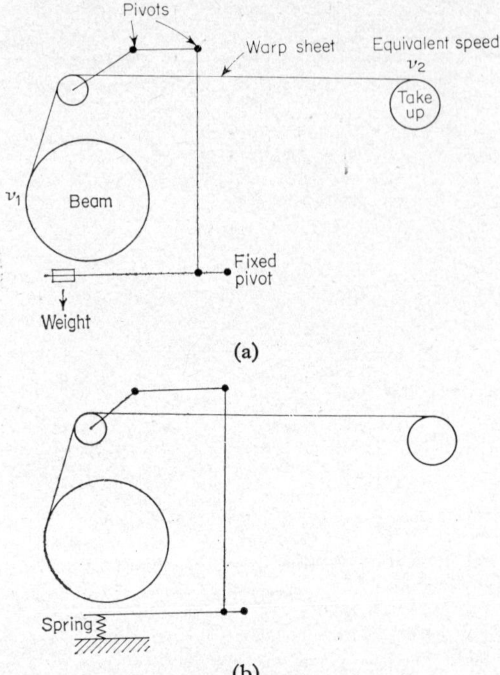

FIG. 7.8. Weight and spring positioning of floating beam in let-off control.

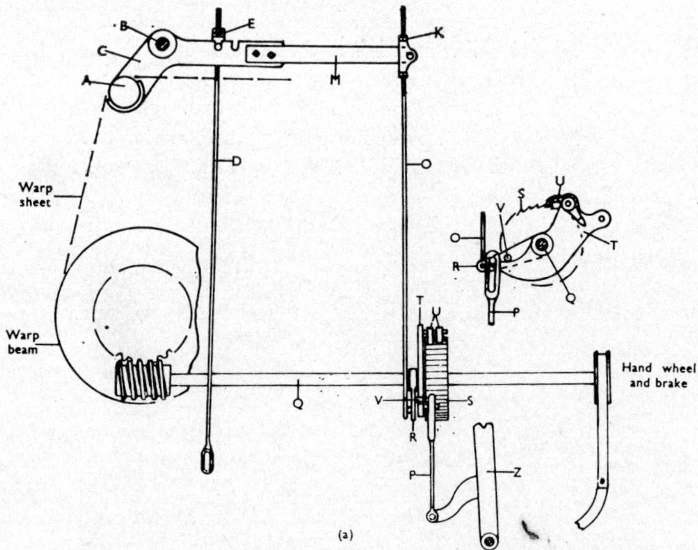

FIG. 7.9. Hattersley light standard let-off motion.

position of the roller and a fixed point produces a speed change in the ratchet wheel S directly proportional to this difference. The detailed action is as follows. The lever P is moved down from the position shown, and then up again through a fixed distance during each loom cycle. In so doing the levers R and T are moved by a distance which depends on the position of lever O. If the stud in lever R is at the top of the slot in lever P it will be moved the maximum distance; if the stud is at the bottom of the slot, R will not be moved at all. Levers O and R return to their original position as lever P moves back again due to the weight acting on the floating roller A. Thus when the warp has been over-fed from the warp beam, floating roller A will move up and lever R will move towards the bottom of the slot where the movement transmitted from P to R is very small. The movement of lever R is conveyed to lever T which actuates the vernier ratchet U so that gear S moves round a distance proportional to the position of the floating beam. The rotation of S is conveyed to the warp beam by means

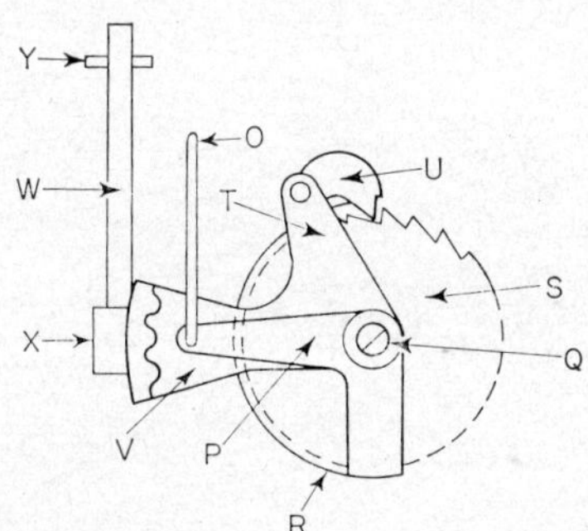

Fig. 7.10. Shield and ratchet let-off mechanism.

of the worm gear drive shown in Fig. 7.9 so that the warp beam is rotated at a speed depending on the position of the floating beam.

An alternative form of ratchet speed control is shown in Fig. 7.10. In this system the ratchets are moved forwards and backwards through a fixed distance during each loom cycle, but are prevented from engaging all the teeth by the shield R. This shield is once more positioned from the rod O which records the position of the floating roller A. The principle of both these systems is clearly the same.

The warp knitting machine let-off mechanism shown in Fig. 7.11 differs considerably in principle. The surface speed of the warp sheet is kept constant at a required value which is fixed by the speed at which disc C is rotated. The speed of the warp sheet is determined by winding a thread S from a separate bobbin round disc V, and then passing it round the warp beam. As the warp beam rotates, it pulls the thread with it, and so rotates disc V at a speed proportional to the speed of the warp sheet. (The end of the thread coming from the warp beam is usually fed to one of the knitting needles for removal after it has fulfilled its task, and it thus finds its way into the cloth.)

The speeds of discs C and V are then compared, and if a difference exists,

o

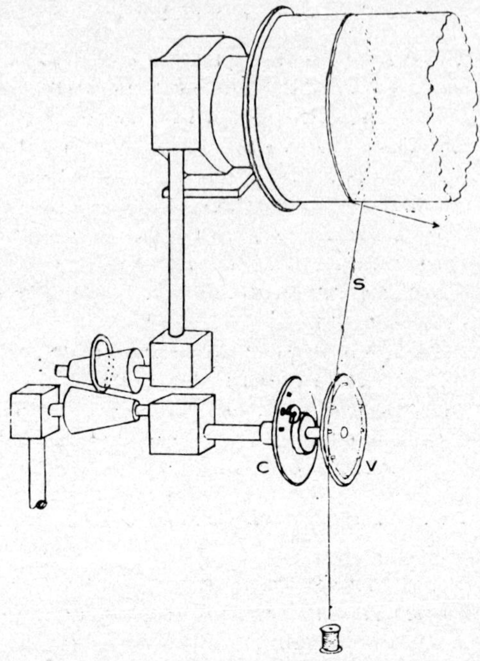

FIG. 7.11. Principle of warp knitting machine positive let-off control.
(Reproduced from W. F. Paling, *Warp Knitting Technology*, Harlequin Press.)

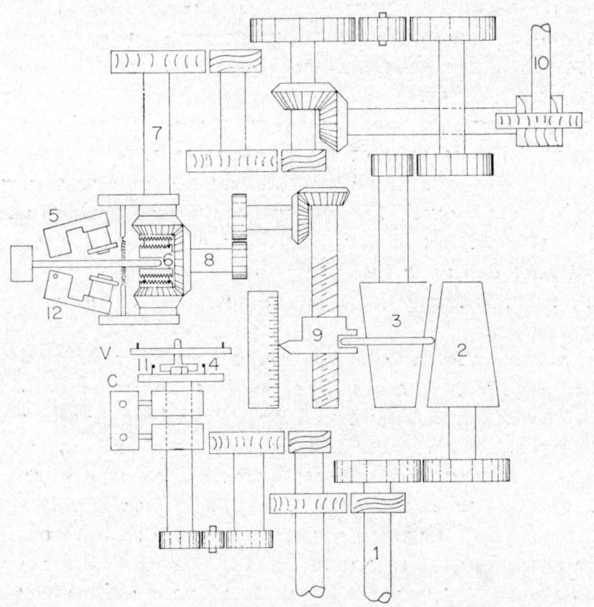

FIG. 7.12. Details of warp knitting machine positive let-off control.

the speed of the warp beam is altered. Thus if disc V moves faster than disc C it will gradually move across the latter until the electrical contacts 4 (see Fig. 7.12) are closed. These contacts are in a circuit with solenoid 5 so that when the contacts are closed the solenoid moves the dog teeth on gear 6 into position, and gear 6 now rotates with shaft 7, on which it is loose until locked to the shaft by the dog teeth. As a result, gear 8 rotates, thus driving the screw 9 in such a direction that the small ring on the two conical pulleys is moved to a position where the driven pulley 3 of the pair rotates at a slower speed. The driving pulley 2 is driven from the main drive shaft 1 which also drives the fixed-speed disc C through a gear train. The warp beam 10 is driven through a gear train by the driven pulley, and its speed therefore decreases until the two discs are once more running at the same speed. If the driven disc is going too slowly it will gradually move back relative to disc C until the contacts 11 are closed. This activates solenoid 12 which results in screw 9 being driven in the opposite direction and the warp beam is speeded up.

The warp knitting device, while having an on-off character, still basically controls the speed of the warp beam in a similar way to the loom let-off devices shown in Figs. 7.9 and 7.10. The detection and comparison sections of the two devices are very different. In the warp knitting machine the speed of the warp sheet is measured, compared with a fixed value, and the error is used to control the speed. In the loom let-off mechanism this is impossible because the required value of the speed is unknown. Instead the path length between the warp beam and the cloth is measured. A change in this length is a change in the integral of the difference between the actual and required let-off speeds. A controller in which the actuating section is fed with a signal which is the integral of the difference between the required and measured values of the controlled variable is known, for obvious reasons, as an integral controller, and such controllers have the useful property that on average the controlled and measured variables must be the same. The loom warp let-off mechanism will therefore control the warp beam speed very accurately on average. The disadvantage of integral controllers is that they cannot be used for controlling rapid changes in the controlled variable without introducing a tendency to hunt. The use of ratchet and wormwheel drives on the let-off mechanism, however, restricts the rate of adjustment of the feed rate so that hunting is not considered to be a problem with such mechanisms.

While on average, the values of the required and actual speeds must be the same this does not imply that the integral of the difference must be zero. In fact, it can be shown that the integral depends on the gain of the system, and is directly proportional to the position of the floating roller; it follows, therefore, that the position of the roller will vary with the gain of the system. The gain varies during the running of the loom due to the change in warp beam diameter. The same action on the ratchets will produce a larger change in warp beam surface speed when it is full, compared with the speed when empty, simply due to the change in warp beam diameter. It follows, therefore, that the gain of the system changes, and hence, that the position of the floating beam will alter with changes in warp beam diameter. This is not of great importance on the dead-weight loading systems shown in Figs. 7.8a and 7.9 since, as has already been pointed out, the tension in the warp beam is, at least to a first approximation, independent of the position of the floating roller. If springs, however, are used to load the floating beam, as in Fig. 7.8b, any movement of the floating beam results in an alteration in the warp tension. This is undesirable and compensating mechanisms have to be fitted to spring-loaded let-off mechanisms to maintain a constant gain during the running down of the warp beam, thus preventing any change in tension in the warp sheet. These mechanisms have been developed on empirical lines and their analysis is too complex to deal with in this brief survey. The difficulties encountered, however, show the kind of problems that have to be overcome when the detector does not directly measure the quantity it is desired to control, but some other quantity related to it. This difficulty is of basic importance in the control of sliver thickness in drafting, which will be considered in the next section.

7.7   THE CONTROL OF SLIVER THICKNESS IN DRAFTING AND CARDING

An ideal system of controlling sliver thickness in drafting would involve measuring the thickness of the sliver leaving the machine, comparing this measured thickness with the required thickness, and then altering the draft ratio accordingly. Thus, what is essentially a closed-loop system would be used. However, altering the draft ratio has the effect of changing the number of fibre ends in the cross-section of the sliver leaving the machine, and a measure of sliver thickness is given by the integral of this number over the mean fibre length. Thus, because this integration has to be performed, there is a time lag (at least equal to the time taken for the sliver to move through a distance equal to the mean fibre length) between the instant the draft ratio mechanism is activated and the detection of the effect of the resulting change in sliver thickness. We have already seen that any appreciable time lag in a closed-loop system results in the system having a natural tendency to hunt. Hunting, in a drafting system, is to be avoided as it would lead to a particularly undesirable form of irregularity in the sliver. As a result the only commercially available methods of controlling drafting irregularities are based on open-loop control systems, which cannot hunt.

Fig. 7.13 shows the basic parts of the Raper autoleveller. Before entering the drafting zone the input sliver passes between two grooved rollers. The axis of the bottom roller is fixed, and the roller is driven at a constant speed. The upper roller is carried on a weighted lever. The sliver is compressed into the groove between the rollers, by the pressure of the top roller. Thus if the thickness of the input sliver increases, the top roller will rise, and vice versa. The movement of the top roller is thus a measure of the thickness of the ingoing sliver. The small movement of the top roller is magnified by the lever system shown in Fig. 7.13, the roller movement being measured from a reference position based on the required ingoing sliver thickness. This magnified movement is used to move the belt along the two conical pulleys. The back rollers are driven by these variable-speed pulleys so that the draft is varied according to the thickness of the input sliver.

In the simple system described so far, however, no account has been taken of the time required for the sliver to move from the rollers to the drafting zone. Consequently the draft ratio would be changed before the section of sliver whose thickness was measured reached the drafting zone. A delay mechanism must therefore be incorporated to ensure that the draft is changed only when the measured part of the sliver reaches the drafting zone.

One form of delay mechanism is shown in Fig. 7.13, and consists of a rotating drum with a series of slots cut in its surface. A steel rod which protrudes from the surface of the drum is placed in each of the slots. The magnified movement of the top roller is stored in these rods by the setting arm moving the rods a distance proportional to the movement of the top roller. The wheel rotates, in a

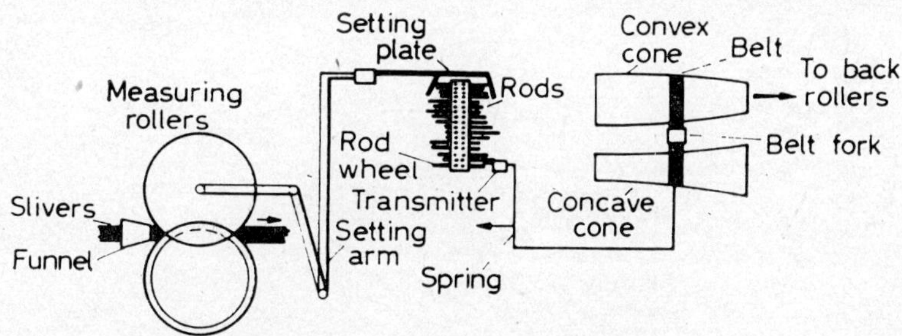

FIG. 7.13. Raper autoleveller.
(Reproduced from the *Textile Manufacturer*.)

plane at right-angles to the plane of the paper, at such a speed that the time taken for a rod to pass from the setting plate to the transmitting lever is equal to the required delay time. The rods move the transmitting lever, and hence the belt on the two pulleys, by the same amount as the rods were moved originally by the setting plates.

Many different forms of autoleveller have been devised. The measurement can be made electrically by the use of capacitance or β-ray meters. The delay factor can also be obtained by a variety of electronic means, the final speed variation can depend on variable-speed gearing, such as the PIV gear (Fig. 2.17), or by the use of variable-speed electric motors, etc., but the basic principles of all these devices are the same. It should be noted that an autoleveller of this kind will only correct the thickness variations of the outgoing sliver which are due directly to variations in input thickness. Any irregularity caused by the drafting mechanism itself cannot be corrected for. This is basically true of any open-loop control but it is of particular importance in this case because other causes of irregularity exist. The open loop is still used, however, since only very recently has a closed-loop control been developed successfully for this control application.

Fig. 7.14, on the other hand, shows the autocount system—used to control the count of the sliver produced by a woollen card—which is a closed-loop control. The count of the sliver is determined by measuring the amount of light from a standard light source at point A which will pass through the carded web. The amount of light is compared with a standard, and if a difference occurs, the variable-speed gear which drives the doffer is actuated, the amount of the adjustment in speed being a function of the difference between the measured amount of light and the standard. If the web is too thick the amount of light passing through it will be too small, and the variable-speed gear is so adjusted as to speed up the doffer and reduce the thickness of the web. The web will be reduced in thickness at point B, where the doffer takes the fibres from the swift, and it will take some time (approximately 10 seconds) for the effect of this change

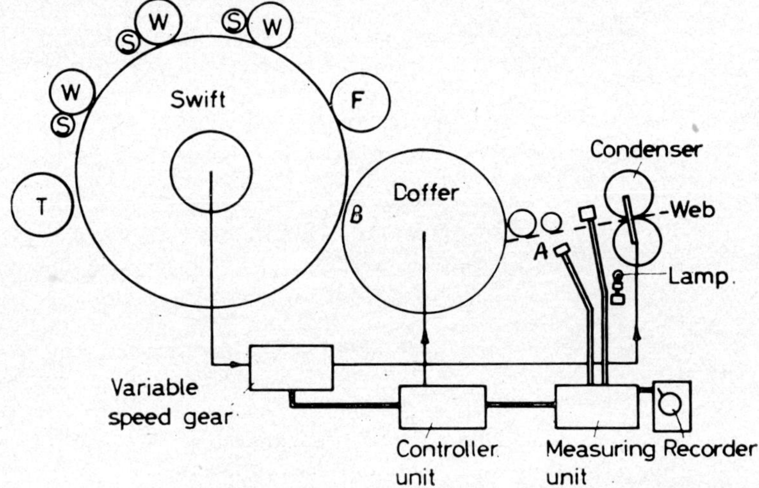

FIG. 7.14. Autocount controller for woollen carding machines.

in speed to be detected at the measuring point A. As we have pointed out before, such a delay in measurement will result in the system having a tendency to hunt. It is not possible in this instance to use an alternative open-loop control, so another method of overcoming this difficulty has been devised. The gain of the controller was made very small, and in addition an integral controller was used. In other words, the speed of the doffer was not adjusted by an amount proportional to the difference between the measured and required thickness, but by the integral of this difference. Such an integral controller with low gain produces a sluggish control action which has no tendency to hunt and which will give a highly accurate long-term control of count. The sluggishness of the action, however, results in the short-term irregularities of the sliver being unaffected by the controller.

7.8 THE DETECTION OF FULL OR EMPTY PACKAGES AND THEIR REPLACEMENT

The detection of full packages in winding, twofold twisting, and similar processes has become extremely common in recent years, since it has been found that considerable savings in waste are possible if the packages all have the same length of yarn on them. These detectors are usually operated from the feed rollers, and record their number of revolutions; the machine being stopped when the required number of revolutions has been made.

A different principle is used to detect when the bins and hoppers of cards and scutchers have been overfilled. Such a mechanism is essential in an integrated blow-room–card-room arrangement where, for example, the card hopper is fed continuously from the scutcher. A photo-transistor is placed in one wall of the hopper opposite a light placed in the opposing wall, the light and transistor being placed at the maximum height to which the bin is to be filled. If the bin is overfilled the beam of light is cut off and the resistance of the phototransistor is thus increased. This results in an electronic circuit closing a relay which stops the previous machine in the flow line. Such electrical analogues of the mechanical switching circuits described in the previous chapter are essential where the distance from the detecting point to the point of application of the effect is large.

The detection of empty packages, in warping or drafting machinery, for example, is usually carried out by using yarn detector-stop mechanisms. When the yarn has run out, or nearly run out, the machine stops and the package is replaced. A particularly interesting example of such mechanisms is that used to detect the existence of an empty pirn in a loom shuttle. In addition to detecting an empty pirn, these mechanisms have been developed so that they replace the empty pirn by a full one without stopping the machine, such replacement being of the right colour or type of yarn when different yarns are being used in a multiple-box machine.

There are several types of detection unit, all of which are relatively simple. The most commonly used detection unit consists of two prongs connected to an electrical circuit which will be described later. These prongs are inserted into the shuttle when it is stationary. The pirn has a metal insert on its shaft which is usually covered by the yarn on the pirn. When the yarn is nearly used up the metal is uncovered and the two prongs complete the circuit shown in Fig. 7.15, through the metal insert.

The circuit shown is for a four-one box loom in which there are four boxes with up to four different kinds of yarn in shuttles on one side of the loom. One of these shuttles is fired, as selected by the box mechanism, into a single box at the other end of the loom where the pirn is tested for the amount of yarn left, and if it is nearly empty the pirn in this box is subsequently changed. By closing the circuit shown in bold lines in Fig. 7.15a the solenoid is activated, thus completing the circuit shown in bold in Fig. 7.15b. There are four such circuits in a four-box loom, all of which are attached to the feeler terminals, though each is connected to a different switch in the contact box, and has a separate magnet box in its circuit. The contact box switches are operated from the box mechanism chain so that if the shuttle from box 2 has been fired, switch 2 in the contact box is closed. Only the circuit and solenoid which have a closed contact in the contact box are activated when the feeler detects the near-empty pirn. As a result, only circuit number 2 will be closed and the magnet box in circuit 2 will be activated. Each magnet box has a lever which releases a pirn from a different magazine of pirns. As soon as the pirn appropriate to shuttle box 2 has been released the shuttle-changing mechanism operates and changes the pirn in this shuttle.

This general description of the circuit assumes that the pirn can be detected and changed in the same loom cycle. This is not feasible mechanically, and the

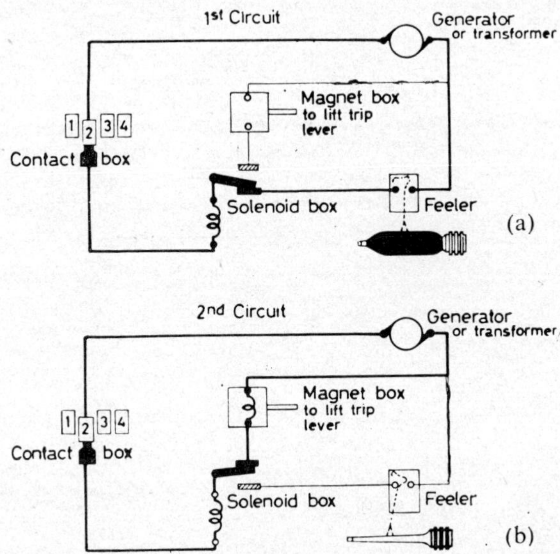

FIG. 7.15. Replacement of pirn selection mechanism.

circuit has been modified slightly to enable the pirn to be changed when next it arrives in the single box. The modification consists simply of a catch in the solenoid box which keeps the switch closed, as in Fig. 7.15b, once it has been closed by the presence of an empty pirn. This catch is only released when the pirn is finally changed. The action of the circuit is thus as follows. When shuttle 2 arrives in the single box and the near-empty pirn is detected, switch 2 in the selector being closed, solenoid 2 and its switch are closed, and latched closed. The timing of the changing mechanism is such that no change of pirn takes place due to this selection. Shuttle 2 is fired back to the four-box side. Subsequently, shuttle 2 will be selected once more for firing from the four-box side. As soon as this happens switch 2 in the contact box is closed but as switch 2 in the solenoid box is already closed, pirn 2 will be selected from the magazine and dropped into position, so that as soon as shuttle 2 arrives in the single-box end it is immediately ready for changing. As soon as the change occurs the latch on the solenoid switch is lifted mechanically and the system is ready for the next detection of an empty pirn.

Such a control mechanism clearly requires a fairly elaborate decision-making section, and its complexities are increased considerably in the automatic pick-and-pick loom, where four shuttle boxes are used on both sides of the loom. Automatic pirn changing involves other fairly complex mechanical detection mechanisms to ensure, for example, that the shuttle is correctly positioned in the single box when the pirn is about to be changed, and to ensure that the free end of the yarn from the pirn is positioned correctly, and held, when the pirn is changed. For details of these mechanisms the reader is referred to the specialised textbooks on weaving mechanisms at the end of the previous chapter.

7.9 THE CONTROL OF TEMPERATURE

Many textile processes (e.g. dyeing) have to be carried out in conditions in which the temperature must be controlled fairly strictly. Consequently, mechanisms for temperature control are common in the textile industry. They can broadly be classified into two basic types, depending on whether the heating system is electrical or whether it is operated by steam.

When the heating system is electrical it is usual for the control mechanism to be electrical also. A very common electrical on-off controller consists of a normal mercury thermometer placed in the liquid, the temperature of which is to be measured. One part of the controlling circuit for the electric heating leads to the thermometer bulb, and the other ends in a pin which lies along the line of the mercury thread, and whose free end is at the point which the mercury reaches when the required temperature is reached. When this temperature is reached the mercury closes the gap in the circuit which, through a relay, switches the heater off. When the temperature drops, the mercury drops, the circuit is broken, and the heater switched on again.

When the heating system is operated by steam, the control system is usually pneumatic, and the basis of such a system is normally the Bourdon gauge. The bulb of the gauge is placed in the liquid to be heated. As the temperature rises the liquid which fills the bulb expands and causes the arms of the Bourdon gauge to move apart. One arm is kept fixed and the other moves a small vane towards, or away from, a small pneumatic jet. When the vane is close to the jet the air pressure is high; as it moves away the pressure drops. As a result the air pressure is a function of the temperature of the bath. This air is fed to the top section of the control valve shown in Fig. 7.16. When the pressure is high the diaphragm is depressed and the valve closes, decreasing the flow of steam to the heater. Pneumatic controllers are extremely robust, cheap, and reliable, but they cannot normally be used in other textile applications because they are basically slow acting.

The use of such closed-loop controllers makes it possible for any process which is temperature controlled to be programmed. For example, the temperature in a dyeing vessel can be arranged to be maintained at one temperature for a fixed time, followed by a second temperature for a further period. The set point in the controller in such programmed processes will usually be moved by

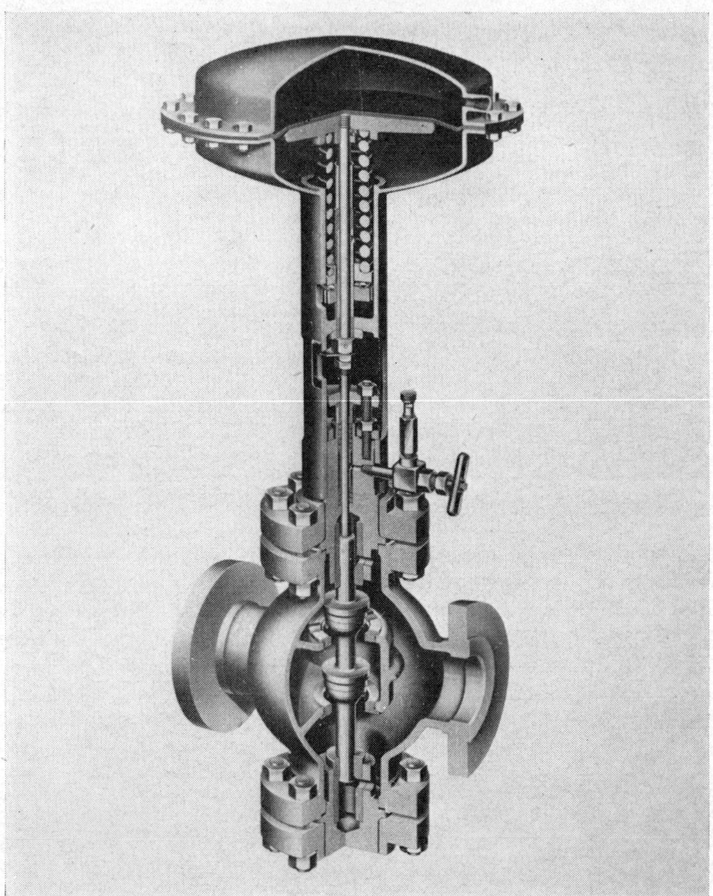

FIG. 7.16. Diaphragm valve (by courtesy of J. Blakebrough and Sons Ltd).

a cam driven by a clock mechanism. Fig. 7.17 shows an example of such equipment used for controlling and programming the temperature in a heat setting-chamber. The control used is pneumatic.

Such a programmed controller is an example of the combination of a selection and a control mechanism. Such systems have been considerably developed in the chemical industry and, in fact, completely automatic factories have been

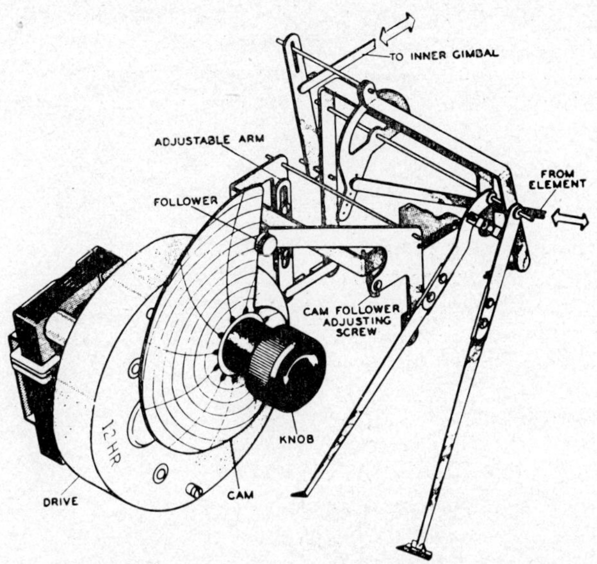

FIG. 7.17. Programmed temperature controller
(by courtesy of the Mason-Neilan Regulator Company).

built. Such developments will almost certainly occur in some branches of textile processing in the future.

Suggested Further Reading

A. H. NISSAN (Editor). *Textile Engineering Processes*. Chapter 13. Butterworths, 1959.
A. J. YOUNG. *An Introduction to Process Control System Design*. Longmans, 1955.

INDEX

Acceleration
  cam follower, 83, 87
  chain drive, 40
  linkage mechanism, 112, 113
  sley, 112, 114
Addendum, 44, 54
Amonton's Law, 29
Autocount, 190
Autoleveller, 31, 188, 189

Bearded needles, 117, 119, 123
Bearings
  Air, 10
  ball, 12
  cylinder, 12, 16
  journal, 7, 17
  linear, 17
  magnetic, 12
  roller, 12
  self-aligning, 15
  sintered, 3
  spindle, 16
Belt slip, 25
Bevel gears, 42
Binary information store, 137, 138
Binary numbers, 136
Bottom board, 148, 149, 151, 153
Box motions, 164
Brinton drum, 142, 163, 166
Builder motions, 129

Cams
  conjugate, 79
  cycloidal, 98, 100, 101
  cylindrical, 78

disc, 75, 76, 77
  end, 76
  face groove, 78, 127
  harmonic, 98, 100, 101
  knitting, 91
  linear, 80, 101
  parabolic, 84, 98, 100, 101
  picking, 2, 101
Card, 39, 40, 190
Centre shed
  dobby, 152
  jacquard, 149
Chain drive, 38
Circular pitch, 44
Closed loop control, 173, 190, 194
Closed shed
  dobby, 152
  jacquard, 149
Combing mechanisms, 128
Cone
  pulleys, 30, 70
  winder, 21
Conjugate
  cams, 79
  gears, 48
Controller
  gain, 176, 182
  hunting, 176, 188
  feed back (closed), 173
  feed forward (open), 173
  on-off, 172, 175
Cycloidal
  cam shapes, 98, 100, 101
  gears, 55
Cylinder bearings, 12, 16

Dedendum, 44
Diametral pitch, 44, 45, 53
Diaphragm valve, 193
Differential gears, 68, 70, 72
Disc
    cams, 75, 76, 77
    tensioner, 180
Dobby
    Knowles, 154, 164
    Leeming, 157
    positive, 153
    Staubli, 159, 167
    V-, 152
Doubling frame, 129
Draft
    gear, 64
    ratio, 63, 64
Drafting
    rollers, 22
    mechanism, 38
Dwell, 79, 114, 115, 116

Elasticity, 6, 88, 101
Elements, 1
End cam, 76
Epicyclic gears, 64, 72, 129

Face groove cam, 78, 127
False twister, 12
Flyer spinner, 31, 32, 70
Follower
    acceleration of, 83, 87
    force on, 88
    knife edge, 76, 89, 91
    motion of, 80, 81, 85, 99, 101, 104
    mushroom, 76
    roller, 76, 94, 106
    vibration of, 97, 98, 103, 105
F.N.F.
    guide bar mechanism, 120–3
    let-off mechanism, 185

F.N.F. needle bar mechanism, 120–
    123
Friction, 29, 91
    drives, 19 et seq.

Gate tensioner, 180
Gear(s)
    bevel, 42, 72
    change, 63, 74
    cycloidal, 55
    differential, 68, 70, 72
    efficiency of, 56, 59, 73
    epicyclic, 64, 72, 129
    helical, 42
    idler, 62
    involute, 49 et seq.
    worm, 42, 56, 74
    skew, 42
    spiral bevel, 42
    spur, 42, 43, 56, 63, 73, 74
    trains, 61, 74
Generation, 52
Geneva mechanisms, 166 et seq.
Guide bars, 123, 134

Half hose machine, 134
Hattersley
    dobby, 152
    let-off motion, 183
Headstock gear train, 63
Heald shaft threading, 143
Helical gears, 42
Helix properties, 57
Hertz, 88
Hrönes, 117, 118
Hydraulic actuators, 130
Hydrodynamic lubrication, 4, 7, 8, 9
Hydroelastic lubrication, 7
Hydrostatic lubrication, 4, 10

Indexers, 166 et seq.

Information
  loops, 174
  store, 132, 134, 137
Involute, 49 et seq.

Jacquard, 133, 147 et seq.
  cylinder, 147, 149, 150, 166

Knife-edge follower, 76, 89, 91
Knitting
  action, 93, 117, 119
  bars, 117, 119, 123
  cams, 91 et seq.
  machine, circular jacquard, 141
    flat bed, 77, 127, 137
    warp, 119, 120, 134, 185
    weft, 93, 141
  needle, 91, 142
Knowles dobby, 154, 164
Kopp variator, 33

Linkage chain, 108
Links, driving, 108
Loom
  box motion, 164
  label, 140
  picker stick, 1, 2, 102
  picking cams, 2, 101, 105
  pirn selection, 192
  shuttle, 1, 2, 102, 113
  sley, 112, 114
  tappet, 144

Machine, 1
Magazine, 192
Magnetic bearing, 12
Mangles, 22
Mechanism, 1
Module, 45, 54
Mushroom follower, 76

Needle bar, 117, 119, 123
Negative let-off, 180
Nelson, 117, 118

Open loop control, 188
Open shed dobby, 152
Open shed jacquard, 151

Pairs,
  sliding, 2, 16
  turning, 1, 2, 7
Paper tape, 132, 139, 160
Parabolic cams, 84, 98, 100, 101
Pawl-and-ratchet, 124, 166, 184, 185
Photo-electric scanner, 138, 163
Pirn detector, 191
Pitch circle, 44, 45
P.I.V. gear, 37
Planetary gear, 64
Poisson's ratio, 89
Positive feed, 31
Pressure angle, 51, 94
Programmer, 194
P.T.F.E., 3
Pulse, 97
Punched card, 132, 138, 140

Rack, 52, 75
Raper Autoleveller, 31, 188
Restraint, 1, 15
Roller
  bearing, 12
  chain, 38
  follower, 76, 94, 106
Rolling contact, 6
Rotational speed, 19

Schweiter, 21
Scroll indexer, 170
Scutcher, 191
Shield and ratchet mechanism, 126, 185

Silent chain, 40
Simple harmonic motion, 80, 112
Skew gears, 42
Sliding pairs, 2, 16, 18
Sliver thickness, 188
Slurcock drive, 127
Solenoid, 187
Sommerfeld number, 9
Speed ratio, 20, 30, 40, 43, 58, 62, 68,
  124
Spindle
  bearings, 16
  tapes, 24
Spiral bevel gears, 42
Spur gearing, 42, 43, 56, 63, 73, 74
Staubli dobby, 159, 167
Structure, 108

Take-up roller, 74
Temperature controller, 193
Tension
  belt, 24
  centrifugal, 29
  control
    closed loop, 181
    disc, 180
    gate, 180
    negative, 180
  limiting, 26
  warp, 183
Thrust, 14
Time delay, 172, 176
Timing belts, 40
To-and-fro motion, 75
Transmission of power, 25

Tooth
  height, 53
  movement, 73
  shape, 46
Turning points, 109
Twist rollers, 72

Variable speed, 30
Velocity, 82, 85, 99, 104
Velocity gradients, 5
Vernier ratchet, 125
Vibration, 97, 103, 105
V-belt, 27, 35

Warp
  breakage detector, 177
  let-off, 183
  tension, 183
Warp knitting
  cams, 79
  let-off, 35, 185
  linkage mechanisms, 120, 123
Wear, 2
  effects of, 14, 49, 101
Weft breakage detector, 178
Weft knitting
  action, 93
  cams, 78, 91
  selection, 133, 137, 141
Worm gears, 56

Yarn
  breakage detector, 178
  friction, 29
  tension, 179

Printed in Great Britain by Butler & Tanner Ltd., Frome and London